MAILING ADDRESS
Maritime Studies Research Unit
Memorial University of Newfoundland
St. John's, Newfoundland A1C 5S7, Canada

RESEARCH IN MARITIME HISTORY
NO. 39

TAIWANESE DISTANT-WATER FISHERIES IN SOUTHEAST ASIA, 1936-1977

Henry T. Chen

International Maritime Economic History Association

St. John's, Newfoundland
2009

ISSN 1188-3928
ISBN 978-0-9738934-9-6

Research in Maritime History is available free of charge to members of the International Maritime Economic History Association. The price to others is US $25 per copy, plus US $5 postage and handling.

Back issues of *Research in Maritime History* are available:

No. 1 (1991)	David M. Williams and Andrew P. White (comps.), *A Select Bibliography of British and Irish University Theses about Maritime History, 1792-1990*
No. 2 (1992)	Lewis R. Fischer (ed.), *From Wheel House to Counting House: Essays in Maritime Business History in Honour of Professor Peter Neville Davies*
No. 3 (1992)	Lewis R. Fischer and Walter Minchinton (eds.), *People of the Northern Seas*
No. 4 (1993)	Simon Ville (ed.), *Shipbuilding in the United Kingdom in the Nineteenth Century: A Regional Approach*
No. 5 (1993)	Peter N. Davies (ed.), *The Diary of John Holt*
No. 6 (1994)	Simon P. Ville and David M. Williams (eds.), *Management, Finance and Industrial Relations in Maritime Industries: Essays in International Maritime and Business History*
No. 7 (1994)	Lewis R. Fischer (ed.), *The Market for Seamen in the Age of Sail*
No. 8 (1995)	Gordon Read and Michael Stammers (comps.), *Guide to the Records of Merseyside Maritime Museum, Volume 1*
No. 9 (1995)	Frank Broeze (ed.), *Maritime History at the Crossroads: A Critical Review of Recent Historiography*
No. 10 (1996)	Nancy Redmayne Ross (ed.), *The Diary of a Maritimer, 1816-1901: The Life and Times of Joseph Salter*
No. 11 (1997)	Faye Margaret Kert, *Prize and Prejudice: Privateering and Naval Prize in Atlantic Canada in the War of 1812*
No. 12 (1997)	Malcolm Tull, *A Community Enterprise: The History of the Port of Fremantle, 1897 to 1997*

No. 13 (1997) Paul C. van Royen, Jaap R. Bruijn and Jan Lucassen, *"Those Emblems of Hell"? European Sailors and the Maritime Labour Market, 1570-1870*

No. 14 (1998) David J. Starkey and Gelina Harlaftis (eds.), *Global Markets: The Internationalization of The Sea Transport Industries Since 1850*

No. 15 (1998) Olaf Uwe Janzen (ed.), *Merchant Organization and Maritime Trade in the North Atlantic, 1660-1815*

No. 16 (1999) Lewis R. Fischer and Adrian Jarvis (eds.), *Harbours and Havens: Essays in Port History in Honour of Gordon Jackson*

No. 17 (1999) Dawn Littler, *Guide to the Records of Merseyside Maritime Museum, Volume 2*

No. 18 (2000) Lars U. Scholl (comp.), *Merchants and Mariners: Selected Maritime Writings of David M. Williams*

No. 19 (2000) Peter N. Davies, *The Trade Makers: Elder Dempster in West Africa, 1852-1972, 1973-1989*

No. 20 (2001) Anthony B. Dickinson and Chesley W. Sanger, *Norwegian Whaling in Newfoundland: The Aquaforte Station and the Ellefsen Family, 1902-1908*

No. 21 (2001) Poul Holm, Tim D. Smith and David J. Starkey (eds.), *The Exploited Seas: New Directions for Marine Environmental History*

No. 22 (2002) Gordon Boyce and Richard Gorski (eds.), *Resources and Infrastructures in the Maritime Economy, 1500-2000*

No. 23 (2002) Frank Broeze, *The Globalisation of the Oceans: Containerisation from the 1950s to the Present*

No. 24 (2003) Robin Craig, *British Tramp Shipping, 1750-1914*

No. 25 (2003) James Reveley, *Registering Interest: Waterfront Labour Relations in New Zealand, 1953 to 2000*

No. 26 (2003) Adrian Jarvis, *In Troubled Times: The Port of Liverpool, 1905-1938*

No. 27 (2004) Lars U. Scholl and Merja-Liisa Hinkkanen (comps.), *Sail and Steam: Selected Maritime Writings of Yrjö Kaukiainen*

No. 28 (2004) Gelina Harlaftis and Carmel Vassallo (eds.), *New Directions in Mediterranean Maritime History*

No. 29 (2005) Gordon Jackson, *The British Whaling Trade*

No. 30 (2005) Lewis Johnman and Hugh Murphy, *Scott Lithgow: Déjà vu All Over Again! The Rise and Fall of a Shipbuilding Company*

No. 31 (2006) David Gleicher, *The Rescue of the Third Class on the Titanic: A Revisionist History*

No. 32 (2006) Stig Tenold, *Tankers in Trouble: Norwegian Shipping and the Crisis of the 1970s and 1980s*

No. 33 (2007) Torsten Feys, Lewis R. Fischer, Stéphane Hoste and Stephan Vanfraechem, *Maritime Transport and Migration: The Connections between Maritime and Migration Networks*

No. 34 (2007) A.B. Dickinson, *Seal Fisheries of the Falkland Islands and Dependencies: An Historical Review*

No. 35 (2007) Tapio Bergholm, Lewis R. Fischer and M. Elisabetta Tonizzi (eds.), *Making Global and Local Connections: Historical Perspectives on Ports*

No. 36 (2008) Mark C. Hunter, *Policing the Seas: Anglo-American Relations and the Equatorial Atlantic, 1819-1865*

No. 37 (2008) Lewis R. Fischer and Even Lange (eds.), *International Merchant Shipping in the Nineteenth and Twentieth Centuries: The Comparative Dimension*

No. 38 (2008) Adrian Jarvis and Robert Lee (eds.), *Trade, Migration and Urban Networks in Port Cities, c.1640-1940*

Research in Maritime History would like to thank Memorial University of Newfoundland for their generous financial assistance in support of this volume.

Table of Contents

List of Tables

List of Tables

List of Illustrations

Notes on the Sources

Note on the Sources

Interviews, meeting records of the Kaohsiung Fishing Boat Commercial Guild [*Gaosyong Yuye Shangye Tongye Gonghuei*] (KFBCG) and other unpublished materials, as well as in-house periodicals that circulated within Taiwan's fishing industry, were the three most important sources for my research. Without them, I could not have presented a comprehensive picture of the history of Kaohsiung's fishing industry.

Oral History and Fieldwork

Most of the older Taiwanese fishers are either illiterate or semi-illiterate; the lack of any formal education prevented them from producing any written works about their lives or the culture and society of their fishing communities. Therefore, living in the fishing communities and interviewing elderly fishers and their family members was an important methodology. The oral histories I collected during my fieldwork effectively filled in the gaps in the documentary record, enabling me to hear the voices of old fishers and to reconstruct the ethnohistory of these fishing communities. Information collected from the interviews and my fieldwork experiences in Kaohsiung has been used extensively in the book. Information about the informants can be found in appendix 1.

Guild Records and Other Unpublished Materials

The meeting records of the KFBCG were also used extensively to describe and analyze the interactions among the fishing industry, government, military and the ancillary industries. The KFBCG was officially established in 1965. Prior to that date, the Guild had existed for several years as the Kaohsiung Pair-Trawler Owners Club [*Gaosyongshih Shoucaowangye Lianyihuei*], or KPTOC). The KPTOC, however, was an informal group, and its existence had not been approved by the government. As a result, it did not have a fixed headquarters nor was it allowed to recruit permanent members. Vessel owners who attended the club meetings took turns in the chair. The early club records show that besides trawler owners, longliner owners also attended. They met in a restaurant once a month to discuss fisheries-related issues. The records of their discussions were all handwritten.

With the government's permission, the club was formally reorganized as a guild with an office in the Gushan District. The Guild members included both pair-trawler and longliner owners. After the KFBCG was officially established, vessel owners held meetings more frequently. These meetings took three different forms: meetings of supervisors [*Lishihhuei*], combined meetings of supervisors and directors [*Lijianshih Liansi Hueiyi*] and member representatives' conferences [*Hueiyuan Daibiao Dahuei*]. Besides regular meetings, vessel owners held special meetings when necessary. Their regular meetings produced a huge volume of records. Some were handwritten, and others were typed and printed. They were still kept in the Guild office (now in the Cianjhen District) when I conducted my fieldwork in 2002. Most of files clearly indicate the date, venue and title of the meeting.

Since most of the meeting titles were similar and constant, they can be classified into four major types:

Type 1: KPTOC, "1st Meeting Records of the KPTOC in 1962" [*Gaoxyongshih Shoucaowangye Lianyihuei 51 Niandu Diyicih Hueiyi*], 24 February 1962.

Type 2: KFBCG, "4th Meeting Records of the 6th Cohort Directors" [*Dilioujie Disihcih Lishih Hueiyijilu*], 19 August 1976.

Type 3: KFBCG, "19th Combined Meeting Records of the 5th Cohort Directors and Supervisors" [*Diwujie Dishihjioucih Lijianshih Liansi Hueiyi Jilu*], 22 January 1976.

Type 4: KFBCG, "1st Conference Records of the 2nd Cohort Member Representatives" [*Benhuei Dierjie Diyicih Hueiyuan Daibiao Dahuei*], 20 January 1967.

To show readers the source of information, the titles and dates of meetings are clearly specified in the footnotes. But since the formal titles of the meetings do not actually provide information about their substance, I do not list the individual meetings in the bibliography.

In addition to the Guild records, some of my informants, especially Huang Sianchih, the former head of the Fisheries Bureau, Cai Wunyu, a retired politician from Kaohsiung, and Chu Pingyi, a research fellow at the Academia Sinica, kindly offered me numerous significant unpublished documents. These unpublished materials are included in the bibliography.

Periodicals

The use of news and industry reports enables an examination of the long-term patterns and trends in the business environment of Taiwan's fishing industry. There were three major fisheries periodicals published in postwar Taiwan: *Fisheries Tribune* ([*Yuye Luntan*], 1961-1974), *China Fisheries News* ([*Jhong-*

guo Yuye Sinwen], 1956-1974) and *Chinese Fisheries Weekly* ([*Jhonghua Yuye Zhoukan*], 1975-1997). The *Fisheries Tribune* used to be a weekly periodical, but due to financial difficulties it began to publish monthly after No. 23. The *Fisheries Tribune* provided in-depth news analysis of the fishing industry; served as an effective communication channel between the fisheries authorities and the industry; publicized fisheries policies on behalf of the government; and conveyed the needs and demands of the fishing industry to the government.

China Fisheries News and *Chinese Fisheries Weekly* are periodicals that fall somewhere between journals and newspapers. They were printed in the form of a news sheet, and most of their authors remained anonymous. Like journals, however, they were issued regularly, and the contents of the weekly news reports were about more than current affairs in the fishing industry. Like the *Fisheries Tribune*, they also built a bridge between the fisheries authorities and the fishing industry. The government used these periodicals as a political instrument to convey official ideology, while the fishing industry publicized its loyalty to the government in their pages. In chapter six I mention the *China Fisheries Monthly* [*Jhongguo Shueichan Yuekan*] once. This monthly was devoted solely to marine science.

Footnotes, Tables and Figures

To avoid clutter, all titles in the notes, and the sources to tables and figures, are given in English, regardless of the language in which they were published. Readers interested in finding the original sources are urged to consult the bibliography, where the materials are divided by language group (Chinese, Japanese and English). The bibliography also contains the Chinese and/or Japanese titles for works in those languages.

About the Author

HENRY T. CHEN <dr.henry.chen@gmail.com> is a Research Fellow at the China Research Centre (CRC) at the University of Technology in Sydney, New South Wales. He obtained his PhD in Asian Studies from Murdoch University in 2007, and worked as a Research Assistant at the Asia Research Centre for a project on the "Evolution and Development of the Taiwanese Offshore Tuna Fishery, 1913-2006," which was sponsored by the History of Marine Animal Populations (HMAP).

His research interests and special expertise lie mainly in the areas of politics and the maritime history of Taiwan; the culture and history of China; and classical Chinese literature. Henry is multilingual and able to conduct research in Chinese, classical Chinese, Japanese and Taiwanese. As an Australian academic, however, his scholarly papers are all in English. Thus far, Dr. Chen has published widely in Canadian, Australian, French and Dutch journals. Henry is also a translator and interpreter between English and Chinese (Professional level). His T&I proficiency is officially accredited by the National Accreditation Authority for Translators and Interpreters (NAATI).

He is presently engaged in a research project on the "International Distribution of Wealth from Pacific Tuna Industries: A Value-Chain Analysis of Case Studies from Pacific Island Countries," which explores the proportions of total retail value accruing to each group, and the historical, economic, cultural and political factors behind each group's positions in the chains and the relative proportion of wealth they obtain.

Acknowledgements

I would like to thank my parents and my two brothers for their comfort, encouragement and financial and moral support which enabled me to undertake this research. I owe them an enormous debt of gratitude.

I also have to express my heartfelt thanks to Professor James F. Warren and Associate Professor Malcolm T. Tull who over the past few years have offered important advice about research strategies and facilitated my personal and professional growth. I admire their scholarship, and their kindness is deeply appreciated.

I also wish to extend my thanks to the Kaohsiung Fishing Boat Commercial Guild. When I conducted my fieldwork in Kaohsiung (December 2001-July 2002 and January-February 2003), the Guild not only generously lent me its records but also introduced me to important informants. I would like to extend my thanks to key informants, such as Wu Tianreui, Wang Hejia, Lin Changren, Pan Sanguang, Gu Tingfang, Chen Shengli, Cai Wun'yu, Syu Yisin and the many other local people who granted me interviews. They were not only generous with their time but also with many important details. Their life histories and personal insights have made a tremendous contribution to this research. Without their help, this book would have lacked much of the flavour they were able to provide.

Introduction

Although there are a large number of important historical and ethnographic studies of fishing nations like Japan and the United Kingdom, this is not the case for Taiwan. To begin to rectify this situation, this book traces the historical development of Taiwan's fishing industry in Southeast Asia from 1936 to 1977. Because of its location, Taiwan was one of the most important forward supply bases for the Japanese fishing industry in the *Nanyo* (Southern Ocean) during the colonial era (1895-1945). Although World War II and the eventual Japanese defeat had a devastating impact on Taiwan's fishing industry, it soon recovered through postwar governmental policies and financial assistance from international agencies. In the 1950s, Taiwanese pair-trawlers began to expand beyond the Taiwan Strait, first along the southern Chinese coast and then southward to the coasts of Vietnam, Indonesia and finally even northern Australia. At the same time, offshore longliners began to sail southward from Kaohsiung (see figure 0.1) to develop long-distance fisheries between the Philippines and the Chinese province of Hainan.

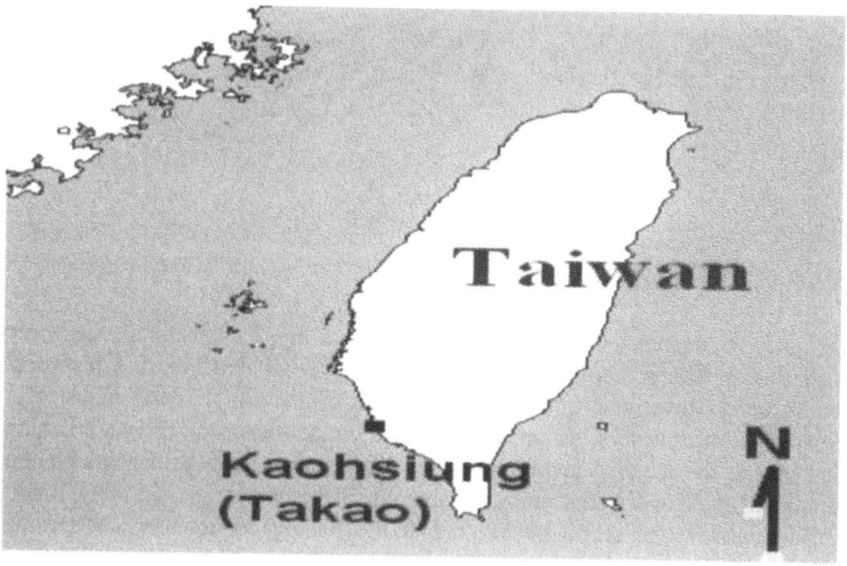

Figure 0.1: Geographical Location of Kaohsiung (Takao)

Source: Courtesy of the author.

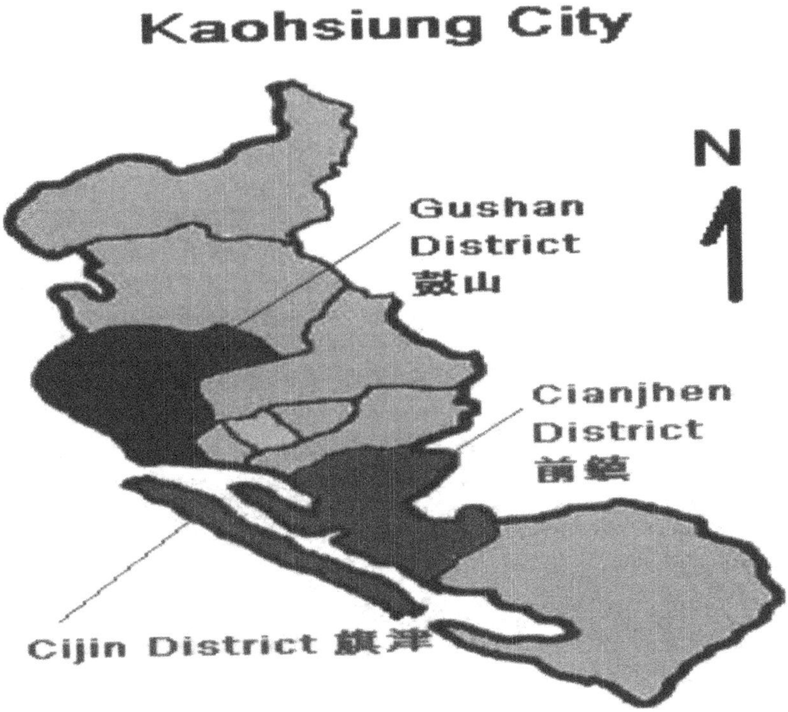

Figure 0.2: Location of Gushan, Cijin and Cianjhen

Source: See figure 1.

To understand the rise and decline of the Taiwanese distant-water fisheries, I will proceed at three levels. First, I will examine the historical background, government policy and the birth of Taiwan's fishing industry in Southeast Asian waters. Second, I will focus on the history of Taiwan's fishing industry at the national level to describe and analyze the relationships among the industry, government, military and ancillary industries. Finally, I will further narrow the scope by looking at the fishing communities, observing management, the daily lives of fishers and aspects of their culture. Because Kaohsiung, which was called Takao in the colonial era, was the most important base for the Taiwanese fleets that fished off Southeast Asia, I will limit this study to that port, which also includes the fishing ports of Gushan, Cianjhen and several small places in the Cijin district (see figure 0.2). Gushan, where the authorities constructed a large market, storage facilities and petroleum tanks, was the cradle of the Taiwanese Southeast Asian fishing industry during the colonial period and remained the most important port for offshore fishing until 1967. The Cijin district, on the other hand, was less developed; before the war its fishing activities were devoted to aquaculture and the coastal fishery. But when several offshore harbours were built there after the war, local fishers began to venture offshore. The creation of the fishing port at Cianjhen in 1967 was a landmark event, and the port soon became the centre of Taiwan's deep-sea fishing industry.

I will limit the time frame of this study to the period 1936-1977 for several important reasons. I begin in 1936 because that was when Kobayashi Seizō, a Japanese Navy Reserve Admiral, was appointed Taiwan's Governor-General and launched the "Marching Southwards Policy" (*Nanshin Seisaku*), which aimed to control the *Nanyo* through both military and economic measures. To this end, the Taiwanese fishing fleets were encouraged to expand their activities southward. In the process, they also collected intelligence for the Japanese navy and boosted Japanese prestige in Southeast Asia. Forty-one years later, in 1977, the United States claimed a 200-mile Exclusive Economic Zone (EEZ), which encouraged many other nations to follow suit. As a result, Taiwan's traditional offshore fishing grounds were blocked from expansion by a number of coastal nations.

It is also important at the outset to establish the difference between the terms "distant-water fishery" and "offshore fishery," which in the case of Taiwan cannot be readily defined without taking account of the historical development of its fishing industry. This occurred in two phases. The first lasted from the 1910s to the 1960s. In the early twentieth century, industrial fishing, using longliners and pair-trawlers, spread from Japan to Taiwan. Most vessels were owned by companies and fished in the South China Sea and the waters off Southeast Asia. Vessel owners called this a "distant-water fishery" because it was carried on "far away" by the standards of the time. The second phase covered the period thereafter. In the 1960s, distant-water vessels left the seas off Southeast Asia and ventured into the Indian and Pacific oceans. A new wave of longliners that had previously fished in Taiwanese coastal waters now filled the vacuum in the vacated grounds off Southeast Asia. These fishers simply called their new fishing pattern the "offshore fishery." In other words, the distant-water and offshore fisheries in this book both refer to Taiwanese fishing activities in Southeast Asia in the years between the 1930s and the late 1970s.

In the first part of the book I describe and analyze the development of Taiwanese colonial and postwar fishing policies. I introduce the administrative systems and their contributions in detail and describe how they affected the diffusion of fishing methods during the colonial period and how they utilized international financial aid in the postwar era. In the process, I explain how and why Kaohsiung was fashioned into the major supply base for Taiwan's Southeast Asian fishing industry in the years 1936-1977.

In the second part of the book I examine the interactions between the fishing industry and government, the military and ancillary industries, underscoring the importance of the political and social aspects of the industry in the second half of the twentieth century. During that period, the industry was under the strict surveillance of the government and the military. Hence, I will examine the measures that these two bodies adopted to control Taiwan's fisheries and discuss the industry's response. I will also describe the tactics used by vessel owners to influence affairs. To secure their business interests, these owners put considerable personal effort into dealing with shipyards, ice manufacturers and other ancillary industries. They partnered with those who were cooperative and boycotted those they regarded as potential enemies. I examine how vessel owners dealt with threats from other industries as well as with people of lower socio-economic status, such as gear suppliers, stevedores and security guards. I then analyze the vessel owners' behavioural characteristics and culture.

Against this background, in the third part of the book I compare and contrast the cultures and management styles of the trawling and longline companies through case studies of specific firms with a special emphasis on the influence of modern fishing technology on the expansion off Southeast Asia. The Taiwanese have fished these waters for many decades; this enables me to make some detailed comparisons both over time and among the various grounds. I also discuss the fishers' roles, duties and onboard politics, and explore their feelings about working at sea for long periods of time.

Different types of fisheries offered various terms of employment to fishers. I examine the interactions between fishing companies and fishermen, and compare job opportunities and welfare practices; in the process, I also observe the influence of ethnicity and culture on the development of Kaohsiung's fishing industry as well as the kinds of fisheries-related activities people undertook. The book provides an in-depth examination of the fishers' daily lives and the culture of their communities, and describes and analyzes their individual and communal responses to major changes in the industry. The final section examines the politics of the implementation of the 200-mile EEZ and its profound impact on the Taiwanese offshore fishing industry.

In short, this is both a regional and local study which not only highlights Taiwan's fisheries development but also enhances our understanding of the daily lives and culture of Taiwanese fishers. It makes two especially important contributions to the scholarly literature. First, it underscores the regional significance of the history of the Taiwanese offshore fisheries and provides a detailed study of its development and demise. Second, I hope it will encourage fisheries historians to undertake further investigations of some of the issues raised in the following pages.

Chapter 1

Japan and the Development of Taiwan's Fishing Industry

Taiwan's modern fishing industry was established during the Japanese colonial period (1895-1945). This initial phase of development featured a spectacular increase in productivity and laid a solid foundation for future activities. The purpose of this chapter is to describe and analyze Japan's influence on Taiwan's fishing industry. Initially, I will answer two important questions: why did the colonial government decide to mould Taiwan into one of the principal forward bases for the fishing industry? And what geographic advantages did Taiwan, especially Takao Port (Kaohsiung), possess for the exploitation of the fishing grounds in Southeast Asia? After examining the roles of Taiwan and Takao in the southward expansion of fishing, I will focus on Japan's contribution to the establishment of the fishing industry in Taiwan from a technical perspective and will discuss the range and diffusion of fishing methods that emanated from Japan and the resulting exploitation of the southern fishing grounds. In the final section of the chapter, I focus on the emergence of Takao's fishing industry, investigate the internal and external conditions of its home port and explain why it surpassed all other regional ports to become the centre of the Japanese fishing industry in Southeast Asia.

Marching Southwards Policy and Nanyo Fisheries

The idea of "Southward Development" [*Nanshinron*] was first proposed by Japanese intellectuals at the time of the Meiji Restoration. They asserted that since the economies, societies and cultures of Southeast Asia were still underdeveloped, it was Japan's "duty" to control and civilize the region.[1] But this mission required an outpost, and Taiwan was viewed as the ideal choice. Prior to the formal annexation of the island in 1895, some Japanese had already foreseen the significance of Taiwan's location as an aid to Japan's southward development. Tokutomi Soho, referred to as the "Guide of the Japanese Empire" [*Teikoku Nihon no Kyodosha*], argued that "Taiwan is situated at the

[1]Yano Tōru, *The Genealogy of Southward Development* (Tokyo, 1997), 53-55. Although most of Southeast Asia had been controlled or colonized by Europeans in the Meiji period, some Japanese still believed that God had entrusted the region to Westerners only temporarily and that it would eventually come under Japanese rule.

southern doorway of our country. If we intend to expand the territory of the Japanese Empire southwards, without a doubt the most important thing we need to do is to control this doorway."[2] He also declared that Japan would have to defend its northern frontier before initiating this southward expansion. With Taiwan as a southern outpost, Japanese influence would be ideally placed to penetrate further into Southeast Asia.[3]

This expansionist perspective was regarded as the template for the "Theory Respecting Taiwan as a Southward Fortress" [*Taiwan Nanshin Kyoden Ron*], and it was soon adopted by colonial officials in Taiwan. Katsura Taro, Taiwan's second Governor-General, deemed that Taiwan's geographic position not only enabled Japan to block China's development but also provided an excellent starting point from which to extend Japanese political and commercial influence. Such ideas were implemented gradually after the First World War. When the conflict prevented Europeans from selling industrial products to their overseas colonies, the markets in Southeast Asia were quickly glutted with Japanese goods. Colonial bureaucrats expected Taiwan, as the southern doorway to the burgeoning empire, to play an active role in this southward economic expansion. At that time, censuses of population and land ownership in Taiwan were completed and significant advances made in the production of rice and sugar, which effectively increased governmental revenues. Moreover, Taiwanese military resistance had been suppressed. These conditions enabled the colonial government to implement its southward development plans. This period in Taiwan's history is known as the "Southward Development of the Taisho Period" [*Taisho Nanshinki*].[4]

The colonial government established three shipping lines between Taiwan and Southeast Asia and set up branches of the Bank of Taiwan in Java. In addition, an act entitled "Special Expenditures for Establishing Facilities in South China and the *Nanyo*" [*Nanshi Nanyo Shisetsuhi*] was promulgated. Shimomura Horoshi, the Chief Secretary of Civil Affairs [*Minsei Chokan*], expressed an opinion similar to that of Tokutomi in 1915. He described Taiwan as "a gateway of Japan" and believed that it should become one of the principal outposts of the empire because of its strategic location relative to China and East and Southeast Asia.[5] It is clear that the southward development policy was underway.

---

[2]Gotō Kenichi, *Modern Japan and Southeast Asia* (Tokyo, 2001), 79-80.

[3]Yano, *Genealogy*, 148.

[4]Gotō, *Modern Japan*, 81-82; and Takashi, "Taiwan, Southern China and the Southern Ocean," in Takashi (ed.), *Japan's Involvement in the Southern Issues and the Role of Taiwan* (Nara, 1988), 14-15.

[5]Gotō, *Modern Japan*, 81-82.

In 1919, a system of civil governorship was adopted, and the tradition of appointing governor-generals in Taiwan was abolished. But this important change in the bureaucratic apparatus did not alter the intention to dominate Southeast Asia. The colonial government now turned its attention to investigating the industrial development and economic exploitation of Southeast Asia to pave the way for further control.[6]

Fishing was considered an important economic sector in Japan and was expected to play a major role in southward economic expansion. As early as 1914, a fisheries official, Takayama Itaro, pointed out that the migration of fishers to Southeast Asia could solve the problem of overpopulation in Japan's fishing villages. Moreover, the expansion of fishing grounds could boost Japanese prestige in the region and tap the market for marine products in places such as Singapore. In 1917, the slogan "March Southward! Japanese Fishing Industry!" [*Suisan Nanjin*] was created in Japan.[7]

Several supply bases had been established in Southeast Asia by Japanese fishers even before the economic expansion began. Due to its spatial proximity, the first was Manila, and in 1901 Yamane Yosabee started to fish in the bay. His excellent catches encouraged others, especially from Hiroshima and Okayama, to try their luck.[8] Besides the Philippines, Japanese fishers began to use ports in the Netherlands East Indies to explore nearby marine resources. In 1915, Tamagusuku Tokushuke used Sumatra's fishing ports as supply bases, and six years later Nakamura Itaro used Sabang. Tamakushuku Toku did the same at Manado in the Celebes in 1916. By the mid-1920s two fishing companies, *Kyoei Kumi* and *Daisho Koshi*, operated fisheries-related businesses in Batavia. Singapore became an important supply base from the 1910s; as a free-trade port, it provided Japanese companies with a relatively friendly environment and by 1936 had become one of the most such bases in Southeast Asia. One estimate suggests that 1038 Japanese lived and fished in Singapore at the time.[9] The remarkable success of Japanese fishers in exploiting overseas marine resources also encouraged many others to migrate.

---

[6]*Ibid.*, 82. The investigation included an exploration of marine resources in Southeast Asia. As well, the colonial government ordered the construction of some experimental vessels to exploit new fishing grounds.

[7]*Ibid.*, 48; and Hu Xinghua, *The Development of the Fishing Industry of Taiwan* (Taipei, 1996), http://www.fa.gov.tw/chn/fish_culture/history, accessed 16 March 2005.

[8]Kataoka Chikashi, *Japanese Fisheries in Nanyo* (Tokyo, 1991), 22. In 1900 Tagawa Shintaro became the first Japanese to fish in Manila Bay. Unfortunately, he was unfamiliar with the local fishing grounds and achieved poor results.

[9]*Ibid.*, 48-49, 69, 91 and 93.

An aggressive proposal for the fisheries was put forward in 1934 by Kunishi Kōsuke, an extremely patriotic expert who claimed that although the size of the navy had been strictly limited in 1922 by the Washington Naval Conference, the construction of modern, distant-water fishing fleets could effectively compensate for this deficiency. Since such fleets were equipped with modern telecommunication devices and were suitable for long-distance navigation, he argued, they could also be used for military purposes. Moreover, the fishers could serve as a naval reserve.[10] This plan was adopted by the navy when the international situation became increasingly tense.

In 1936, Kobayashi Seizo, a Japanese Naval Reserve Admiral, was appointed Taiwan's Governor-General. "The Marching Southwards Policy" [*Nanshin Seisaku*] soon became one of his most important strategies. The activities of the navy directly encouraged the southward development of the fishing industry. The navy expected fishing vessels to collect all sorts of intelligence, and Japanese fishers willingly cooperated.[11]

The need to meet Japanese food demands was also an important reason to exploit the Southeast Asian fisheries. By the early 1930s the northern grounds, particularly in the East China Sea, had been gradually depleted, and many believed that they had reached their peak. Yet demand in Japan continued to rise as tensions with China worsened. The supply problem was exacerbated in the late 1930s when Japan suffered from a rice shortage. To solve the problem, fishers were encouraged to operate on comparatively new fishing grounds, such as the Taiwan Strait and the South China Sea. The exploitation of marine resources would not only make a major contribution to the food crisis but also further the creation of an East Asia Co-Prosperity Sphere [*Daitoa Kyoeiken*] (see figure 1.1).[12] Taiwan, and especially Takao, was expected to play a substantial role. Indeed, its position became even more important when Japanese fishers were refused supplies at some bases in Southeast Asia as anti-Japanese sentiment began to rise in the 1930s.

---

[10]*Ibid.*, 259-260.

[11]Gotō, *Modern Japan*, 58 and 94.

[12]*The Fishing Industry of Taiwan* (Taihoku, 1935), 16; *Fisheries Economics, Vol. 1: Fisheries Issues in the South* (Tokyo, 1941), 91-93; and Koshiyoshi Yoshinobu, "The Fishing Industry of Taiwan," in Takemoto Iichirō (ed.), *Series on the Economy of Taiwan* (Taihoku, 1938), 129 and 133-134.

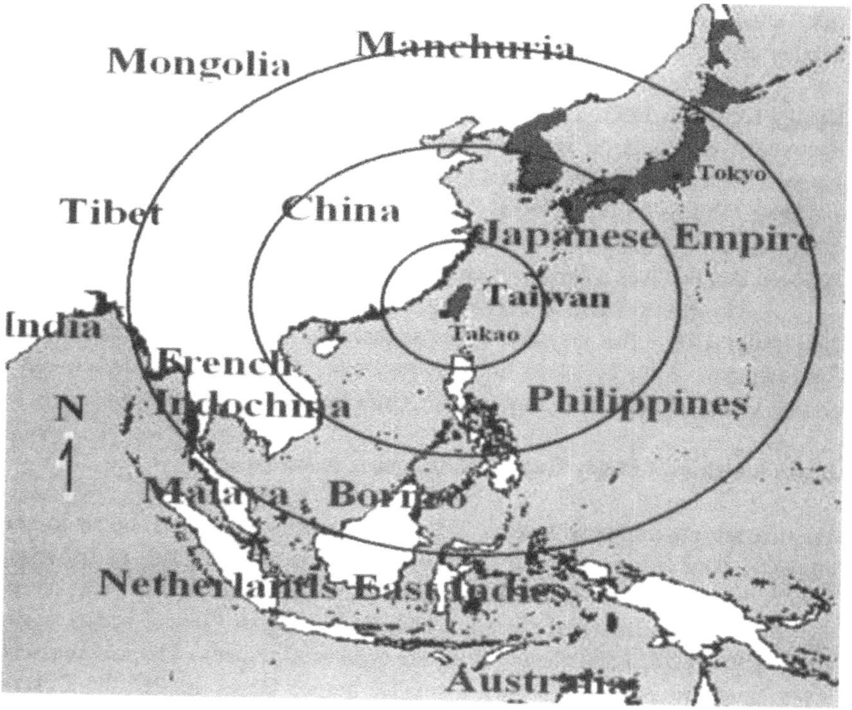

Figure 1.1: Japanese View of Asia with Taiwan at the Centre of the East Asia Co-Prosperity Sphere

Sources: Courtesy of the author based on *A Southern Stronghold, Taiwan* (Tokyo, 1944); and Gotō Kenichi, *Modern Japan and Southeast Asia* (Tokyo, 2001), 81-82.

Japan's economic expansion into Southeast Asia also caused concern among the colonial powers in the region. The government of the Dutch East Indies began to take precautions in the 1930s. Several strict fisheries-related regulations were issued by Dutch authorities to control Japanese migrants engaged in the fisheries, and in 1937 the government formally prohibited foreign vessels from fishing in its territorial waters, a decision which impacted heavily on Japanese distant-water fishers.[13]

The offshore fisheries of British Malaya were dominated by Japanese migrants. During the 1930s numerous Japanese fishing boats used its ports, raising British concerns over security because Malaya hosted one of the most important Royal Navy bases in East Asia. To discourage the Japanese, after February 1937 the colonial government began to issue harsh fisheries-related

[13]*Fisheries Economics, Vol. 1*, 40-41.

rules aimed at creating an unfriendly environment. The policy became even tougher after the outbreak of the Second Sino-Japanese War.[14]

The Commonwealth of the Philippines stopped issuing new licences to foreign fishers in 1933. The deteriorating situation in the late 1930s had a further negative impact on Japanese fishers. The outbreak of the Second Sino-Japanese War in 1937 and Japan's expansionist policies alarmed the Americans and their Filipino hosts. In 1938, the government amended its fisheries regulations to impose greater restrictions on foreign fishing enterprises. The Japanese believed that this was a precursor of tougher measures to follow.[15]

The environment in *Nanyo* became more difficult in the 1930s, especially after 1937. The Japanese believed that this was a result of the Second Sino-Japanese War. To avoid future harassment, many Japanese distant-water fishing vessels shifted their supply bases from Southeast Asia to Takao.

Japan's Fisheries Policy Goals and Business Interests

The marine environment around Taiwan is exceptionally well suited to the propagation of marine life and the development of various kinds of fisheries. Two ocean currents flow along the coasts, and their interactions bring abundant marine resources to the island's shores. The Japan Current brings warm water with a high saline content from the equatorial regions. The main current flows along the east coast, while a smaller branch flows through the Taiwan Strait. The two currents meet in the East China Sea and advance towards Japan, bringing stocks of migratory fish, including mackerel, bonito and tuna. The Littoral Current flows southward along the Chinese coast [*Zhongguo Yanan Liu*], endowing Taiwan with rich resources, such as pomfret, eel, porgy, yellow croaker and cutlass fish (see figure 1.2).[16]

The East China Sea lies north of the Taiwan Strait, covering an area of about 950,000 square kilometres; south of the Strait is the South China Sea, encompassing approximately 206,000 square km. Both are adjacent to Taiwan. Moreover, the main rivers of China, such as the Yangtze and Pearl, carry a prodigious quantity of organic matter to the seas and form an ideal breeding ground for marine species.[17] These two seas, combined with the Taiwan Strait, span an area stretching from northern China to the southern tip of Vietnam. At the beginning of the twentieth century, this was one of the world's largest

---

[14]*Ibid.*, 45-46.

[15]*Ibid.*, 49-52.

[16]*The Development of the Fishing Industry* (Nantou, 1971), 2; and *A Study of the Fishing Industry of Taiwan* (Taipei, 1974), 66-67.

[17]*The Fishing Industry of Taiwan* (Nantou, 1953), 1.

demersal fishing grounds. These marine resources enabled the onshore communities to fish all year, a condition which proved extremely advantageous in the initial stage of the development of Taiwan's fishing industry.

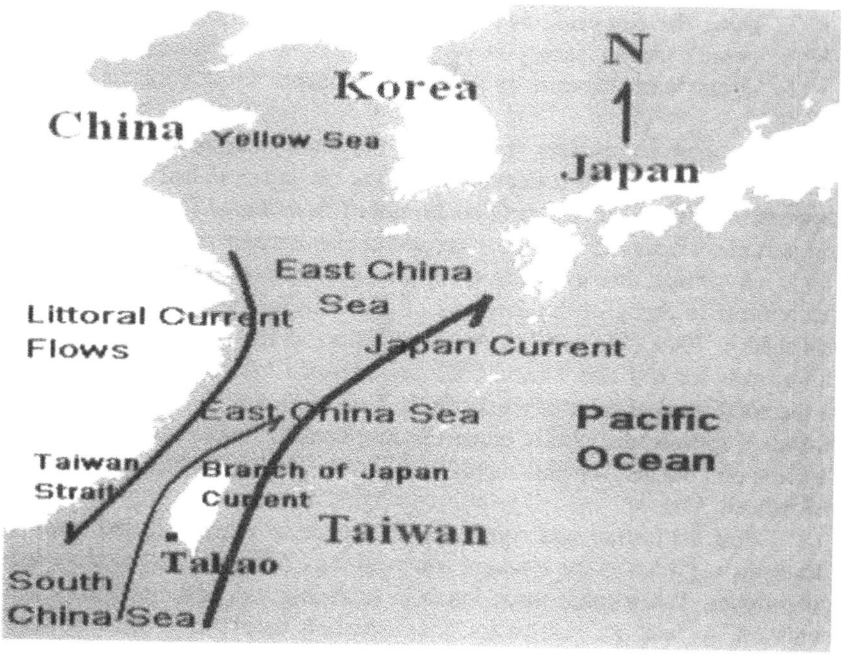

Figure 1.2: Ocean Currents around Taiwan

Source: Courtesy of the author.

Because Taiwanese fishing ports are considerably closer to this major fishing grounds than those in Japan, vessels sailing from Takao could save considerable steaming time, thus lowering operating costs.[18] More important, the use of Takao as a supply base could avoid the boycotts that were increasing elsewhere in Southeast Asia. The colonial government recognized all this and expected that Taiwan, and especially Takao, would play a significant role in the southward expansion of the fishing grounds. These expectations were frequently expressed in official publications and fishing magazines throughout the colonial period. "Industries in Taiwan and Their Leading People" [*Taiwan Jigyōkai to Chūsin Jinbutsu*], published in 1919, pointed out that:

---

[18] *A Brief Introduction to the Agricultural Industries, Vol. 10: Fisheries* (Taipei, 1962), 5; and Zhang Baoshu, *Studies in the Establishment of China's Fisheries* (Taipei, 1952), 20.

> Taiwan is located on the southwestern end of our country [Japan]. Her west coast is separated from southern China by a narrow strait, and the shallow water is suitable for the development of offshore fisheries. The Japan Current flows along the east coast throughout the year. Consequently, the oceanic bonito fishery is very abundant. In the future, the distant water fisheries of Taiwan will flourish.[19]

This observation underscores the fact that the commercial advantages of Taiwan's location were well understood and that the Japanese had foreseen that important offshore fisheries could be developed from there. Takao, as the biggest port in southern Taiwan, was expected to play a major role.[20]

A similar, but even more explicit, perspective can be found in "A Comprehensive Study of the Progress of Takao" [*Yakusin Takao no Zenbō*], published in 1940. The authors described Takao as the doorway of southern Taiwan which could also constrain the progress of China and Southeast Asia. As the southernmost point of the Japanese Empire, it was given the significant mission of spearheading future economic development. Takao became not only a trading port but also a major supply base for the developing fishing industries in Southeast Asia.[21]

Just as Taiwan was considered the southern gateway to the empire, Takao was regarded as the southern entry point to Taiwan. As a transport hub and entrepot, Takao could assist Japan in extending its commercial influence southward. As well, its location could be extremely beneficial to the southward expansion of the Japanese fisheries. Taiwan, and especially Takao, was consistently mentioned in official circles whenever the topic was discussed. Besides government officials and fisheries experts, some leading Japanese fishing companies also realized that Takao could make a major contribution to building their offshore industries. Several major fishing companies and branches were established, and a huge amount of capital was invested in Takao by the Japan Fishing Company [*Nihon Suisan Kabushiki Gaisha*], Takao Seaweed Gathering and Marketing Company [*Takao Kaiso Saishu Hanbai Kabushiki Gaisha*], Takuyo Fishing Company [*Takuyo Suisan Kabushiki-Gaisha*], Rinken Shop

---

19Kamimura Kendou, *Industries in Taiwan and Their Leading People* (Taihoku, 1919), 17.

20During the colonial period Keelung's fishery was stronger than Kaohsiung's. But with respect to the southward development of the industry, Kaohsiung was always mentioned first due to its strategic location and perceived potential. *A Brief History of Takao City, Ten-Year Anniversary* (Takao, 1934), 88.

21Nakayama Kaoru and Katayama Kiyoo, *A Comprehensive Study of the Progress of Takao* (Takao, 1940), 201.

[*Rinken Shouten Shuchosho*] and Takunan Fishing Company [*Takunan Gyogyo Kabushiki-Gaisha*].[22]

It is clear that its geographic advantages led Taiwan to be regarded as the southern gateway to the Japanese Empire. Moreover, Takao was always considered one of the most important supply bases for the exploitation of regional fishing grounds. To use Takao as home port to tap the fisheries resources of Southeast Asia could satisfy both Japan's expansionist goals and civilian business interests. This perspective fostered a consensus among fisheries authorities and fishing companies. As a result, great efforts were made by the colonial government to develop Takao's fishing industry.

Diffusion of Japanese Fishing Methods and Southward Expansion

Before the advent of the colonial era, Taiwanese fishing vessels were either bamboo rafts or tiny boats with limited ranges. During the colonial period, however, new techniques, including the use of single- and pair-trawls, a longline tuna fishery and other innovations, began to transform the industry.[23]

Single-trawl fishing was introduced by the Taiwan Fishing Co. [*Taiwan Gyogyo Kabushiki-Gaisha*] in 1912. Although the results did not meet expectations, the Taiwan Trawling Co. [*Taiwan Tororu Kabushiki-Gaisha*] soon followed. By the mid-1920s, both companies closed due to the economic depression in Taiwan. In 1927, Taiwan's trawl-fishing industry was resuscitated by the Kyodo Fishing Co. [*Kyodo Gyogyo Kabushiki-Gaisha*]. For the next few years the number of single-trawlers was strictly limited under the direct management of the Japanese Ministry of Agriculture and Forestry (MAF [*Norinsho*]). The fishing grounds off Taiwan were divided into two zones. The northern one included the East China Sea and the Taiwan Strait, while the southern comprised the South China Sea. No more than four trawlers were permitted in each area. As a staunch supporter of this fishery, Kyodo was allowed to monopolize the trawl fishery until 1936, when the limit on the number of single-trawlers was replaced by controls over total tonnage. Another firm, the Horai Fishing Co., was soon fishing in the East China Sea and off the coast of South China with the single-trawlers *Tamura-maru*, *Soga-maru*, *Dainichiminato-maru* and *Meiji-maru*. By 1940 eight single-trawlers were based at the Kiryu Fishing Port (Keelung), and two operated from Takao. Eleven more operating in the South China Sea under licences from the MAF used Takao as a supply base (see table 1.1).[24]

---

[22]*Ibid.*, 208-214.

[23]Koshiyoshi, "Fishing Industry," 121-122.

[24]Lin Jinfa, *The History of the Development of Taiwan* (Taihoku, 1936), 865-866; Koshiyoshi, "Fishing Industry" 129; Koshiyoshi Yoshinobu, "The Fishing

Table 1.1
Development of the Single-Trawl Fishery in Taiwan, 1931-1940

	Number of Trawlers	Annual Catch (tons)	Tonnage (per vessel)	Horsepower (per vessel)	Annual Catch (tons per vessel)
1931	4	1,693,784	225.8	574.00	423,441
1932	4	1,581,828	205.0	507.50	370,457
1933	4	2,259,393	205.5	507.50	564,848
1934	4	3,013,056	251.7	553.50	753,264
1935	4	2,962,001	230.8	512.25	740,517
1936	4	3,251,119	230.8	560.25	812,779
1937	5	4,136,154	209.0	572.00	827,230
1938	6.3	6,124,422	267.4	501.90	972,130
1939	7.9	9,181,455	223.3	536.45	1,162,200
1940	8.9	14,930,234	349.1	542.64	1,674,520

Note: The trawl fishery in colonial Taiwan peaked in 1940 and then declined sharply due to the Second World War. The column "Number of Trawlers" is an average for each year.

Source: Naitō Harukichi and Syu Jiwu, *The History of the Fishing Industry in Taiwan* (Taipei, 1957), 14.

Table 1.2
Development of the Pair-Trawl Fishery in Taiwan, 1931-1940

	Number of Vessels	Annual Catch (tons)	Tonnage	Horsepower	Annual Catch (tons per unit)
1931	66	9,824,442	57.10	106.96	388,498
1932	64	8,513,468	54.19	104.00	266,045
1933	64	12,811,449	54.19	104.00	400,357
1934	64	15,492,798	53.40	103.00	484,149
1935	58	22,900,217	63.50	115.90	789,662
1936	80	22,251,552	61.25	110.00	556,289
1937	70	25,726,870	71.28	112.00	735,053
1938	66	27,255,727	76.09	132.85	825,931
1939	84	30,217,786	78.25	137.42	719,470
1940	82	42,219,092	79.39	142.11	1,029,730

Note: The pair-trawl fishery peaked in 1940, then declined due to World War II.

Source: See table 1.1.

Industry in Taiwan II," *Nanyo Fisheries*, III, No. 9 (1937), 15-17; *Fishing Industry of Taiwan*, 15-16; and *A Brief Guide to the Fishing Industry of Takao* (Takao, 1940), 23.

When the pair-trawl fishery began in Taiwan in 1924, the colonial government imposed strict controls to protect the coastal fishing grounds. At first only twenty units were permitted, but the fisheries authority soon granted additional licences: by 1927 there were thirty units and in 1931 fifty units. In 1936, when fisheries management in Taiwan underwent a major reorganization, single- and pair-trawlers comprised about 10,000 tons (see table 1.2).[25]

The main fishing grounds for the pair-trawlers were the East China Sea, Taiwan Strait, South China Sea and Gulf of Tonkin. Initially, most single- and pair-trawlers operated from Kiryu, with comparatively few using Takao. But when the fish stocks in the north began to shrink, the government began to encourage fishers to exploit the southern grounds. To accomplish this, the number of pair-trawlers operating in the north was limited, while no restrictions were placed on their numbers in the south.[26] This measure accelerated the southward expansion of the trawling grounds.

The longline tuna fishery was introduced in 1913 and developed separately in Takao and northern Taiwan. This fishery prospered in the north because it was located close to the main market in Japan. But since the fishing grounds in Southeast Asia were still too far away, longliners continued to work in nearby areas. As a result, fishers in northern Taiwan eventually could not compete with those in Takao, which became the most important port for Taiwan's longliners. By the end of the colonial period, boats from Takao accounted for eighty-one percent of the tuna catch compared to only sixteen percent for Taihoku (Taipei).[27]

The tuna longline, single-trawl and pair-trawl fisheries were all introduced to Takao as early as elsewhere in Taiwan. Takao's single-trawl industry began with only two vessels – a modest investment by the Japan Fishing Company. Nevertheless, after the beginning of the Second Sino-Japanese War in 1937, some trawlers based on the China coast and in Japan chose Taiwan as a supply base. Nine trawlers of the Japan Fishing Company, each of about 500 tons, moved from Hong Kong to Takao, while two trawlers of the Rinken Shop, each of about 600 gross tons, left Shimonoseki for Takao. By 1945, the

---

[25]According to *Fishing Industry of Taiwan* and *Brief Guide to the Fishing Industry*, the *Sokobiki ami* fishery began as early as 1919, which is obviously different than Ichirō's account; see Koshiyoshi, "Fishing Industry," 133-134.

[26]*History of the Fishing Industry in Taiwan,* 14-15; *An Introduction to the Oceanic Fisheries of Our Country* (Tokyo, 1939), 176-177; and Koshiyoshi, "Fishing Industry," 133-134.

[27]Koshiyoshi, "Fishing Industry," 133-134. The tuna longliners came in a variety of sizes. The smallest were less than five tons, while the largest were more than 100. Most popular were medium-sized vessels from ten to twenty tons.

Japan Fishing Company was building a 1000-ton vessel to be deployed in the Gulf of Thailand, the Indian Ocean and along the coast of Africa.[28]

By 1941 twenty-eight units of pair-trawlers were based in Takao, most of which operated in the Taiwan Strait, off Hong Kong and in the Gulf of Tonkin. From 1936 onwards, in order to expand the fishing grounds southward, the colonial government used Hainan as a supply base for its experimental fishing vessel and obtained excellent results off Vietnam. In 1941, twenty-three units of pair-trawlers were under construction, and many believed that the pair-trawl fishery would soon become quite prosperous.[29]

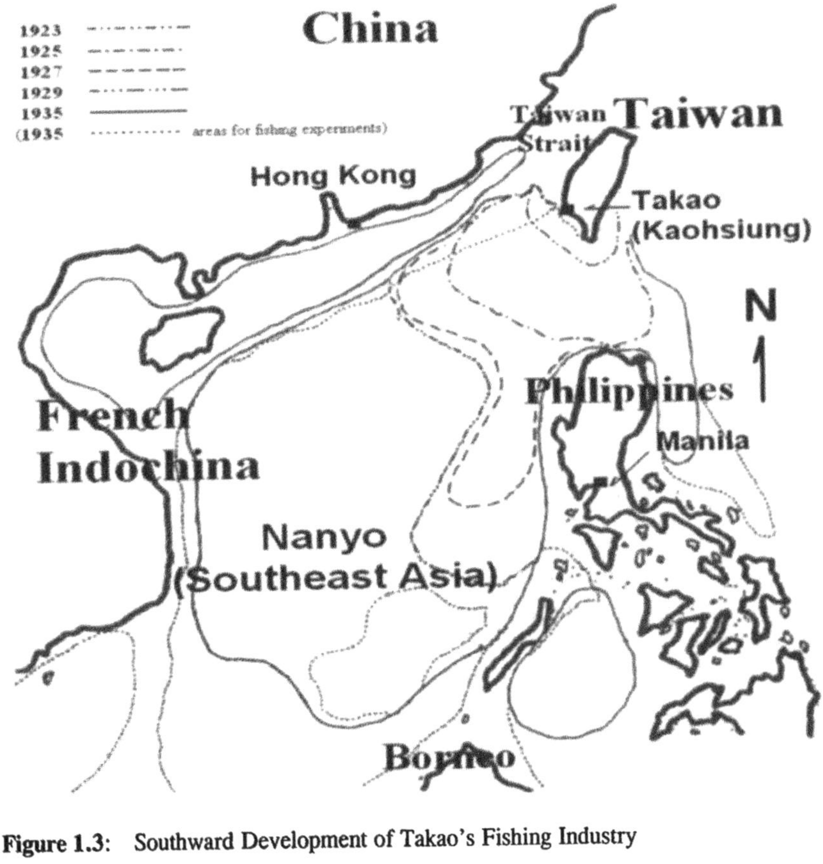

Figure 1.3: Southward Development of Takao's Fishing Industry

Source: Courtesy of the author based on *A Brief Guide to the Fishing Industry of Takao* (Takao, 1940).

[28]*The Fishing Industry of Takao* (Takao, 1941), 6-7.

[29]*Ibid.*, 5; and *The Guide of Takao* (Takao, 1941), 112.

The tuna longline fishery accounted for eighty percent of the motorized vessels in prewar Takao. These were classified into three groups: large vessels of 100 gross tons and 200 horsepower; middle-sized ones of thirty gross tons and sixty horsepower; and small craft of twenty gross tons and forty horsepower. The last category comprised seventy percent of Takao's longliners. Figure 1.3 shows the southward expansion of Takao's fishing industry, while table 1.3 depicts its growth during the period 1938-1940. Catch values grew rapidly, especially in the tuna longline and pair-trawl sectors.[30] By the end of the 1930s fishing had become an important economic activity in Takao.

Table 1.3
Catch Value of the Tuna Longline and Trawl Fisheries in Takao, 1938-1940

	1938 (Yen)	**1939 (Yen)**	**1940 (Yen)**
Tuna Longline Fishery	2,952,759	4,714,939	8,646,761
Pair-Trawl Fishery	807,210	1,693,055	2,253,783
Single-Trawl Fishery	453,857	1,289,116	1,139,371
Total	**4,213,826**	**7,697,110**	**12,039,915**

Source**:** ***The Fishing Industry of Takao*** (Takao, 1941), 2-3.

Construction of Takao Fishing Port and International Markets

In addition to the favourable conditions on the island, there were also local factors that supported the development of Takao's fishing industry. The late nineteenth-century construction of Takao Port was one of the basic keys for the development of the fishing industry, although the original motives for building it were more varied than merely fishing. The rapid growth of the sugar industry in the 1890s required large-scale imports of machinery and facilities to export raw and manufactured sugar. Since it was difficult for the small port to handle this traffic, a decision was made in 1904 to launch an expansion programme. By 1908, the preliminary stages of the project had been completed. Sugar-producing machines could now be transported by train as soon as they arrived at the quay. The link between ocean-going shipping and land transport had a positive influence on the development not only of Takao but also other industries in southern Taiwan, including the fishery.

In 1908 the main railroads in western Taiwan were also connected. Takao now became the southern departure point and railhead for the main railroads; as a result, the port became the economic gateway of southern Taiwan. Rice, sugar, wood and other exports were stored at Takao. But soon the exist-

[30]*Fishing Industry of Takao*, 3-4. These became two of the most important fishing industries in Kaohsiung; their dominance was unchallenged even in the postwar period.

ing infrastructure could no longer handle the traffic, so between 1908 and 1912 the first phase of a major improvement programme was launched. The submerged reefs around the port were removed, two modern harbours established, more rail lines constructed, warehouses and godowns built and an artificial waterway and boat quay constructed for fishing vessels and small boats.

The second phase of construction at Takao led to a remarkable increase in the volume of trade. By 1912, the port handled more than thirty-seven percent of the island's trade, which almost exceeded the volume at Kiryu. Although the second phase was scheduled for completion in 1920, it was not finished until 1937. During this phase the embankment was enlarged, the waterway dredged, the port widened and the number of harbours increased.[31]

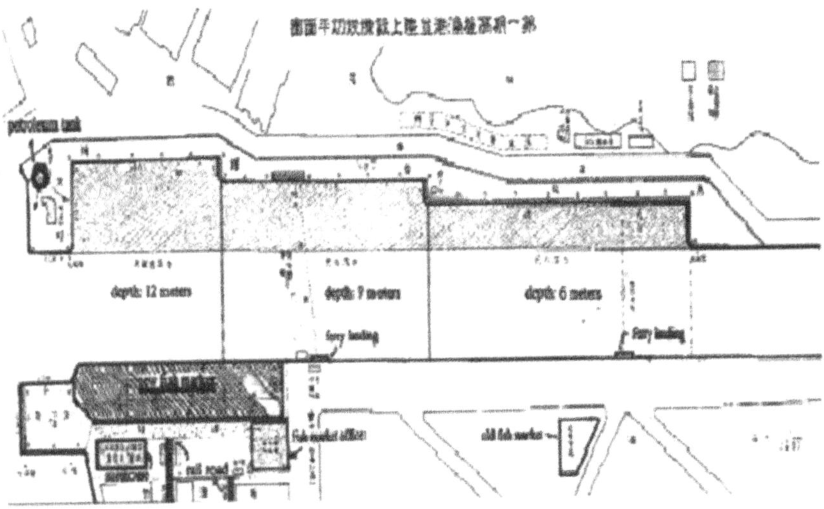

Figure 1.4: Construction of Takao Fishing Port

Source: Courtesy of the author based on Tagame Ichirou, *Takao Fishing Port and its Onshore Facilities* (Takao, 1930).

While these major construction projects were being completed, a special project to extend and modernize the fishing ports was proposed in 1924 by the Transportation Bureau [*Kotsukyoku*]. But it was not approved until April 1926, when Governor-General Kamiyama Mitsunoshin visited Takao. The project included three main aspects: dredging, constructing a protecting em-

[31]The material in the preceding three paragraphs is drawn largely from Zhang Shouzhen and Xu Yinan, *The Past and Present of Hamasen* (Kaohsiung, 1998), 10-15.

bankment at Shōsento Fishing Harbour[32] and building onshore fisheries-related infrastructure (see figure 1.4). The first two projects enabled larger vessels, like the distant-water fishing boats, to anchor, thus directly contributing to the southward development of Takao's fishing industry. The last initiative comprised the construction of a larger fish market, godowns and a petroleum fuel facility. Each made separate but related contributions to the industry. Access to a modern fuel depot supported the motorization of fishing vessels in Takao, which was a major basis for the development of the distant-water fisheries.[33] The large fish market could handle most of the fish products arriving from the Taiwan Strait and Southeast Asia without difficulty.

In addition to the modernization of port infrastructure and the island-wide and international fish markets, the construction of transport networks also made an invaluable contribution to the development of Takao's fishing industry. Markets for Takao's fish were opened in three phases: the establishment of the Takao Fish Market in 1912; combining the Takao market in an organizational chain with other fish markets across Taiwan in 1918; and finally the opening of international fish markets in 1922.

Prior to the establishment of a modern fish market, the fish trade in Takao was conducted by small-scale, itinerant fishmongers, which was disadvantageous to the long-term development of a modern industry. Fortunately, the wholesale and retail situation improved significantly after a modern fish market was established in Takao at the end of 1912 by the Taiwan Sea-Land Property Company ([*Taiwan Kairiku Sangyo Kabushiki-Gaisha*], TSPC). Under the management of TSPC, the fish market became far more efficient and transactions more transparent. As a result, Japanese fishing vessels from Yamagechi, Fukuoka, Kagoshima and Miyazaki began to use Takao as a forward supply base. By 1921, the volume of business in the Takao Fish Market had grown five-fold compared with 1913.[34]

Still, the Takao Fish Market alone could not meet the increased demand. Hence, the opening of other marketing and distribution channels became one of the most pressing goals. But this task could not be accomplished unless a decent transportation network was first built. In 1895, when Japan formally annexed the island, the railroad system was limited to only the northern part of Taiwan. As a result, the transport of fish products was traditionally done by

---

[32]Shōsentō was the first harbour in the Takao Fishing Port.

[33]Tagame Ichirou, *Takao Fishing Port and its Onshore Facilities* (Takao, 1930), 2-10. As a supply base for the southward fishing industries, a modern fuel depot was essential. The tank was made of steel and equipped with a lightning rod and fireproof gear. To prevent an emergency, it was also surrounded by a protective wall. It was located at the entrance to Shōsentō Harbour.

[34]*Ibid.*, 18-19.

sea. In 1899, however, the colonial government began to build railroads in the western part of the island. Lines were developed to Chyoshyū, Taito, Giran and Kaigan. Stations were built gradually. The Takao Train Station had already been constructed by 1900, and by 1908 the main railroad system in western Taiwan had been completely connected so that fish from Takao could be transported by train to northern Taiwan. Two years later, refrigerated railway carriages were introduced, enabling these products to withstand the long journey.[35]

Besides these main railway networks, the sugar companies built private lines deep into the countryside. This was necessary for the transport not only of sugar and related products but also of fish. While the general problem of handling and transporting marine products was solved, the issue of how to transport them from the stations to the local fish markets still remained. Fortunately, the system of roads in western Taiwan and the introduction of trucks for overland transport provided a timely solution by the end of World War I. The establishment of highways can be traced back to 1895 when the Japanese Army Engineering Corps built roads for military purposes. With the subsequent support of the colonial and local governments, roads gradually became more common. After 1919, trucks were frequently employed to convey goods over longer distances, and they became the main means to transport fish products from the stations to the markets.[36]

Because of the continuous growth of Takao's prewar fishing industry, the supply of tuna and sailfish gradually exceeded domestic demand. To stabilize prices, Takao's fishing industry began to look to exports. By the end of 1922 it was proposed that fish products from Takao be transported to Kiryu by train and then shipped by steamer to Japan. Yet this never came into practice until a direct shipping line to Yokohama was established in September 1923. Satō Kanjiro, an important fish broker, took advantage of this opportunity to market fish directly from Takao to Yokohama. But it was difficult to be a pioneer: an unexpected sea disaster on one occasion ruined a shipment of his fish and caused serious financial losses. Moreover, because the containers for fish products consumed a great deal of space, not every shipping company was willing to carry them. These obstacles were gradually overcome, however, and Kanjiro eventually became one of the most successful fish brokers in Tokyo. By 1930, fish products from Taiwan were being exported to all the big cities and coastal towns in Japan, including Tokyo, Yokohama, Nagoya, Osaka, Kobe, Shimonoseki and Moji. By 1930, the value of the fish trade to Japan had reached ¥1,600,000, which accounted for fourteen percent of the annual catch.

---

[35]*Fishing Industry of Takao*, 47-49 and 51-53; and Kamimura, *Industries in Taiwan*, 21-22. In addition to Takao, large markets existed in the other populous towns.

[36]*Fishing Industry of Takao*, 51-53; and Chen Wunshang, *et al.*, *The Ocean in Perspective* (Kaohsiung, 1999), 14.

In addition to Japan, fish from Taiwan could now also be found in the markets of China, Hong Kong and the Japanese colonies of Kantōshū (the southern part of the Liaodong Peninsula in China) and Chosen (Korea).[37]

Japan's influence on the prewar development of Takao's fishing industry can be grouped into three major categories. First, from a macro perspective, Japan's expansionist policies greatly facilitated the growth of the industry. The "Marching Southwards Policy" encouraged distant-water fishing fleets to operate in the waters off Southeast Asia. The outbreak of the Second Sino-Japanese War in 1937 also drove Japan's distant-water fishing fleets to shift their supply bases from Japan, China, Hong Kong and other places to Takao. As a distant-water supply base on the edge of Southeast Asia, Takao's position became increasingly important, especially when anti-Japanese sentiment led the Western powers to boycott Japanese fish products.

Second, the introduction of new fisheries modernized the Takao industry. The tuna longline fishery and the single- and pair-trawl fisheries were promoted by the authorities and civilian fishing companies, and they developed into mainstream industries in the postwar era, generating considerable revenue for Taiwan. Motorized fishing vessels also became increasingly popular. By 1940, there were more than 1200 motorized fishing vessels in Taiwan, and their operations extended from the South China Sea to the Gulf of Tonkin, the Sulu and Celebes seas and the waters off the east coast of Luzon.[38]

Finally, the large-scale exploitation of marine resources in Southeast Asian waters during the colonial era laid a solid foundation for the development of fishing in postwar Taiwan. The prewar operations in Southeast Asia became part of the collective memories and folklore of the Takao fishers. The fishing grounds in Southeast Asia were no longer considered too far away, and they were not afraid to go there in small boats. This positive psychological factor and sense of place at sea proved advantageous to the re-establishment of the Taiwanese fishing industry in postwar Southeast Asia.

---

[37]Tagame, *Takao Fishing Port*, 18-19; and *The Finance of Taiwan's Fishing Industry* (Taihoku, 1930), 15-17.

[38]*Brief Guide to the Fishing Industry*, 109-110.

Chapter 2

The Revival of the Fishing Industry in Postwar Taiwan

Although Taiwan's fishing industry suffered severe damage during World War II, it recovered quickly and achieved outstanding results. The purpose of this chapter is to describe and analyze this revival. Initially, I will provide a concise description of the damage caused by the war to Taiwan's fishing industry and the emergency measures undertaken by the postwar government to aid its recovery. I then examine the key role of the fisheries authorities in postwar Taiwan, describe their functions and show how they worked with one another. After addressing these questions, I explain both the endogenous and exogenous factors that contributed to the remarkable postwar growth. In particular, I draw attention to the fisheries policies and their influence on the development of modern technologies. Besides such internal factors, the revival of the Taiwanese industry benefited immensely from the assistance of the international community.

Emergency Measures for the Recovery of the Taiwanese Fishing Industry

World War II brought severe hardship to Taiwan's fishing industry. Numerous vessels were confiscated for military purposes and were subsequently lost in battle. Much of the rest of the fleet was sunk by American bombers. In short, the fishing industry experienced an unprecedented catastrophe. In 1940 the output from the distant-water fisheries sector was 57,122,328 kilograms. But this dwindled to a mere 68,307 kgs in 1945. The offshore sector contracted from 28,223,394 to 2,599,654 kgs (see tables 2.1 and 2.2).[1] As a result, when the new postwar government took over in 1945, the fishing industry was in a sorry state and was still suffering severely from the devastation of the war years. Emergency measures for the reconstruction of the fishing industry became an immediate priority for the new government.

---

[1]Huang Fucyuan, interview, 20 March 2002; Cai Wun'yu, interview, 7 June 2002; and *The Construction of the Agriculture and Forestry Industries in Taiwan* (Nantou, 1950), 74.

Table 2.1
Decline of the Single-Trawl Fishery in Taiwan, 1940-1944

	Number of Trawlers	Annual Catch (tons)	Tonnage per Vessel	Horsepower per Vessel	Annual Catch per Vessel (tons)
1940	8.9	14,930,234	349.10	542.64	1,674,520
1941	5.4	7,731,454	343.11	667.25	1,431,750
1942	6.1	11,690,422	381.00	644.75	1,916,460
1943	2.4	3,408,708	214.50	558.60	1,420,290
1944	-	646,101	-	-	646,101

Note: The single-trawl fishery in colonial Taiwan peaked in 1940, then declined dramatically due to the impact of World War II. "Number of trawlers" denotes the average number in operation during a year.

Source: Naitō Harukichi and Syu Jiwu, *The History of the Fishing Industry in Taiwan* (Taipei, 1957), 14.

Table 2.2
Decline of the Pair-Trawl Fishery in Taiwan, 1940-1943

	Number of Vessels	Annual Catch (tons)	Tonnage	Horsepower	Annual Catch (per pair)
1940	82	42,219,092	79.390	142.11	1029.73
1941	60	20,618,536	76.977	146.48	687.28
1942	40	8,880,920	65.477	156.32	444.04
1943	5	1,364,893	69.080	101.00	545.93

Note: The pair-trawl fishery in colonial Taiwan also peaked in 1940 and then declined drastically due to World War II.

Source: See table 2.1.

Soon after the Japanese surrender in 1945, personnel were dispatched by the Chinese government to supervise the operation of Taiwan's fishing industry. Numerous fishing companies and related enterprises were taken over and reorganized in a comparatively short time. Fishing Kaisha of Taiwan [*Taiwan Suisan Kabushiki Kaisha*], one of the leading fishing companies in the colonial era, was a good case in point. Most of its vessels had either been sunk or heavily damaged during the hostilities, and many of its branches, ice-manufacturing factories and shipbuilding plants were reduced to ashes. Yet just one year after the end of the war, most of its factories were again in operation, and the damaged vessels had been repaired and were going to sea. Moreover, several new vessels were under construction, and the hope was that forty-six fishing vessels would be in operation by the end of 1946. In November 1945, three months after the end of the conflict, the company's output was a mere twenty-one metric tons, yet by June 1946 output had soared to 393 metric tons. Although this was still well below prewar catch levels, it reflected an extraordinary growth in just seven months.[2]

In addition to taking over the property of fisheries-related enterprises, seventy-one offshore vessels owned by the Japanese were expropriated by the new government. Although most were less than twenty tons, they were still suitable for offshore operations. Hence, they were allocated to local governments to investigate the location of offshore marine resources. In addition, the recruitment of proficient fishers was vitally important to the rehabilitation of the industry. During the colonial period, the majority of vessel mechanics and fishing experts were Japanese, and most had been sent back to Japan. This exodus caused an immediate shortage of skilled labour. To solve the problem, the government reassembled the Taiwanese fishers who left the industry during the war and recruited heavily in China. As well, in 1948 it established training programmes directed by various levels of government. Fishers interested in the coastal fisheries were trained by county governments, while those destined to work on medium- and large-sized vessels were trained by the provincial government in both Keelung (Kiryu) and Kaohsiung. These training programmes produced several hundred offshore and distant-water fishers within two years. This eliminated the worst aspects of the labour shortage and improved the skills of the fishers. The government seemed sufficiently satisfied that it planned not only to continue the programmes but also to improve the skills of longline fishers in order to exploit the tuna resources in Southeast Asia and the Pacific Ocean.[3]

The Fisheries Authorities of Postwar Taiwan

The revival of the offshore fishing industry was the main task of the fisheries authorities in postwar Taiwan. The managerial bureaucracy over the fisheries became extremely complicated in the postwar era, especially after the central government of the Republic of China (ROC) fled to Taiwan in 1949. In this small island nation, semi-official organizations coexisted alongside government departments, and the fisheries authority of the central government functioned side by side with its counterparts at the provincial level. Their respective functions, modes of cooperation and authorities merit careful examination.

---

[2]*Construction of the Agriculture and Forestry Industries*, 197-200.

[3]*Ibid.*, 201; and Jhang Baoshu, *Studies in the Establishment of China's Fisheries* (Taipei, 1952), 32.

The bureaucracy responsible for Taiwanese fisheries development underwent several major changes in the postwar era. The Fishing Industry Division [*Shueichan ke*] of the provincial government was the leading authority immediately after the war. But after 1949, the central and provincial governments on Taiwan were merged. In this unusual political arrangement, the highest fisheries authority became the Department of Agriculture and Forestry of the Ministry of Economic Affairs ([*Jingjibu Nonglinsih*], DAF). As its name suggests, the DAF was concerned mainly with agriculture and forestry; fisheries always had a lower priority.[4]

The postwar fisheries authority of the central government was too small to cater to Taiwan's burgeoning needs. To correct this, in September 1951 the Taiwan Fisheries Production Committee ([*Taiwan Yuye Zengchan Weiyuanhuei*], or TFPC) was founded by the government and its chief ally, the United States. The TFPC included ten members representing various government and business agencies in Taiwan and the US: the ministries of Economic Affairs and National Defence, the provincial government of Taiwan, the Fishermen's Association, the Nonglin Company [*Nonglin Gongsih*], the Council for US Aid, the Economic Cooperation Administration [*Meiguo Jingjihezuo Zongshu*] and the US military consultative group. To emphasize the importance the central government placed on the fisheries, Jheng Daoru, the Minister of Economic Affairs, decided to serve as the "Chief of the Committee" [*Jhuren Weiyuan*].[5]

Following the initial period of postwar growth, in 1953 the TFPC's administrative apparatus was enlarged and reorganized into six administrative units. These changes to a certain extent were due to the fact that Taiwan's fishing industry was developing at such a rapid pace that the old TFPC was inadequate to deal with the problems that emerged. The new structure operated as the *de facto* fisheries authority of the central government for nearly seven years until it was finally dissolved at the end of June 1958.[6] But its abolition did not create a vacuum, since its tasks were now given to a new authority.

In October 1951, immediately after the establishment of the TFPC, the Fishing Industry Division of the provincial government was reorganized as the Fisheries Administration Agency ([*Yuye Guanli Chu*], or FAA). The FAA had the crucial task of administering US aid. The utilization of US aid required permission from various agencies in Taiwan and the US. In the initial stages, the FAA had to make applications to the Council for US Aid or the Sino-

---

[4]*A Brief Introduction to the Fishing Industry of Taiwan Province* (Taipei, 1948), 37-38; and *Fisheries Tribune*, No. 6 (1961), 4-5.

[5]Jhang, *Studies*, 337.

[6]*Ibid.*; *The Fishing Industry in Taiwan* (Nantou, 1953), 15; and *China Fisheries News*, 5 June and 10 July 1958.

American Joint Commission on Rural Reconstruction ([*Nongfuhuei*], or JCRR). Subsequently, all applications were transferred to the US Mutual Security Agency Mission to China [*Ancyuan Fenshu*], which after 1961 was called the Agency for International Development ([*Guoji Kaifa Zongshu*], or AID) for final examination and approval.[7]

If the aid from the US was in the form of a grant, the FAA could usually use its discretion in how to use it. But if it was a loan, the FAA had to re-lend the money to fishing companies or individual fishers. Consequently, from the perspective of the US Mutual Security Agency, the FAA was considered the principal borrower and hence had to repay the loans. Yet from the perspective of the fishing companies or individual fishers, the FAA was actually the creditor. As a key distributor of US aid, the FAA had to ascertain that the fishing companies and fishers were capable of repayment. In this sense, the FAA played a role in bridging the gap between international financial assistance agencies and the fishers and fishing companies in Taiwan.[8]

In 1958, the FAA's apparatus was further enlarged when the TFPC was dissolved. Although the evidence is unclear, it seems reasonable to assume that the purpose was largely to compensate for the abolishing of the latter. Consequently, after 1958 fisheries matters were largely consigned by the central government to the provincial governments. In 1965, the FAA was transformed into the Fisheries Bureau ([*Yuyejyu*], or FB).[9]

In addition to the government departments already mentioned, two other agencies played significant roles in the postwar development of Taiwan's fishing industry. These were the Council for US Aid [*Meiyuan Yunyong Weiyuanhuei*] and the JCRR. The former was established by ROC and the US in July 1947 in Nanjing, the old capital of China. Its main task was to handle the application and utilization of US aid. The Council helped to foster growth in a variety of postwar economic sectors; the fishery was just one among many. The Council was not involved, however, in policy making.[10]

---

[7]*Fishing Industries in Taiwan*, 14-15; *The Bright Future of Taiwan's Fisheries* (Taipei, 1955), 42; *Fisheries Tribune*, No. 10 (1962), 7; and *The Bluebook on US Aid and Achievements of Taiwan's Fishing Industry* (Taipei, 1966), 10.

[8]*Fisheries Tribune*, No. 10 (1962), 8. So that the FAA did not have to worry about bad debts, the task of making and collecting loans was given to two private financial institutions, the Land Bank of Taiwan and the Cooperative Bank.

[9]*China Fisheries News*, 15 May 1958; *The Development of the Fishing Industry* (Nantou, 1971), 3; and *A Study on the Fishing Industry of Taiwan* (Taipei, 1974), 8.

[10]Jhou Siouhuan (ed.), *Documentary Collection on the Joint Commission on Rural Reconstruction, Vol. 1: Government Developments* (Taipei, 1995), 2-3, and *Bluebook*, 9-14.

The JCRR was set up in October 1948, also in Nanjing, by the governments of the ROC and the US. Its main task was to reconstruct rural areas of China using American financial and technical assistance. The fisheries expert attached to the JCRR was Chen Tongbai. In the early years none of the JCRR's plans concerned the fisheries, but Chen soon convinced it to embark on a number of small fisheries-related projects. Over time the JCRR began to attach greater importance to the development of a large-scale fishery, and after 1951 it became heavily involved in the administration of US aid for the construction of offshore fishing vessels. Finally, in 1959 the US Mutual Security Agency consigned its affairs to the JCRR, which made it the most important conduit for US aid to the distant-water fisheries.[11]

As we have seen, many fisheries authorities and fisheries-related organizations amalgamated and were constantly reformed in postwar Taiwan, especially after the relocation of the central government in 1949. This raises the questions of whether this emerging bureaucratic apparatus was sound and whether the various agencies ended up in conflicts. To deal with the second question first, it is surprising, but true, that there were almost no administrative conflicts among the postwar fisheries authorities. The position of the Fisheries Industry Division, as the highest fisheries authority in the first four years of the postwar era, was superseded by the Department of Agriculture and Forestry shortly after 1949. Because the department had many other sectors that it deemed more important, this agency proved inadequate to deal with fisheries affairs. Moreover, it was obvious that the central government did not intend it to be a substantial fisheries authority, primarily because the it still regarded itself as a mainland state and did not want to alter its bureaucratic apparatus to suit an island nation like Taiwan. It was also partly due to the fact that the government did not want to spend additional funds to establish a *bona fide* fisheries unit.

To compensate for the weakness in fisheries management, the TFPC was placed under the Ministry of Economic Affairs. All its members were appointed from government or non-government institutes concerned with the development of the offshore fishing industry; one result was that staffing expenses were minimized. At the same time, the Ministry of Economic Affairs under Jheng became the central government's *de facto* fisheries authority.[12]

---

[11]Siouhuan (ed.), *Documentary Collection, Vol. 1,* 2; Huang Jyunjie (ed.), *Collection of Documents on JCRR* (Taipei, 1991), 130-133; Wu Tianren, *The Biographies of Those Who Engaged in the Early Fishing Industry of Taiwan* (Nantou, 1996), 10; and Huang Jyunjie, *The Reminiscences of the Staff Members of the Sino-American Joint Commission on Rural Reconstruction* (Taipei, 1992), 111-112.

[12]Wu, *Biographies*, 43 and 46. Jheng served as the head of the TFPC's "Projecting and Reviewing Unit" [*Jihua Jiantao Siaozu*], which was set up to provide a blueprint for fisheries development.

Soon after the TFPC was founded, the old Fishing Industry Division of the provincial government was reorganized as the FAA in order to implement the development programmes planned by the TFPC.[13]

In 1958, the central government merged some departments. Although the TFPC had played a crucial role in setting fisheries policy, it was dissolved. At the same time, the FAA was upgraded to take over its affairs.[14] The JCRR was set up by Taiwan and the US to serve as a clearing house for technical and financial interactions between the two nations. Though it was quite influential, the JCRR was only an advisory body. Still, under this complex bureaucratic structure, fisheries management in Taiwan progressed well.

Fisheries Policies and Construction Plans

The exceptional achievements of Taiwan's postwar fishing industry may be ascribed in part to the role of the government. The industry went through seven years of what is often called the "Urgent Reconstruction Period" [*Yuye Chongjianci*] between 1945 and 1952. Several measures to develop the industry were considered and adopted in this initial phase. The first involved the "motorization of fishing vessels and the industrialization of the fisheries," which was implemented by providing loans to fishers. In 1946, the slogan "Revive Oceanic Fisheries" [*Fusing Haiyang Yuye*] was proposed, and four major approaches were taken to fulfil this task: building and repairing vessels to revive production; establishing an effective system for fisheries management; supplying modern fishing gear and equipment to the industry; and training fishers and technicians.[15] These four approaches, although obviously basic, proved to be the proper remedial steps to rebuild Taiwan's fishing industry.

A series of "Regulations Concerning the Distant-Water Fisheries of Taiwan Province" [*Taiwansheng Yuanyang Yuye Guanli Banfa*] were issued to provide a framework for development. These not only provided a legal basis for the establishment of private companies in the distant-water fisheries but also paved the way for the entry of vessels which left China in 1949.[16] These early regulations were an important milestone in the industry's development.

---

[13]Jhang, *Studies*, 336.

[14]*Study on the Fishing Industry of Taiwan*, 8.

[15]Huang Sianchih, "The Past, Present, and Future of Taiwan's Fisheries" (unpublished government ms., n.d.), 1; *A Brief Introduction to the Fishing Industry of Taiwan* (Nantou, 1972), 16; and Hu Singhua, *The Development of the Fishing Industry of Taiwan* (Taipei, 1996), http://www.fa.gov.tw/chn/fish_culture/history/show_article.php?id=29, accessed 6 July 2005.

[16]*Development of the Fishing Industry*.

Increase vessel tonnage was one of the main goals, and it was implemented rigorously through the "Regulations Concerning Single-Trawlers and Pair-Trawlers of Taiwan" [*Taiwansheng Lunchuan Tuowang Jichuan Yiwang Yuchuan Guanli Banfa*], promulgated in August 1948, which imposed minimum limits on the tonnage of fishing vessels and established "fishing exclusion areas" [*Jinyucyu*] to protect the stocks. Under the regulations single-trawlers had to be a minimum of 200 tons, while each pair-trawler had to be between forty and 130 tons. Those that did not meet these requirements were not permitted in the trawl fisheries.

The victory by the Chinese communists led many fishers to relocate to Taiwan. In addition to the state-run fishing companies, by May 1950 eleven private fishing firms, including sixteen single-trawlers and fifty-six pair-trawlers, had relocated. Nonetheless, the total number of trawlers was still only half the prewar level. Most used either Kaohsiung or Keelung as home ports from which to operate in the East China Sea, Taiwan Strait and South China Sea.[17] The extent of these fishing grounds shrank dramatically after the ROC withdrew its troops from the Zhoushan Archipelago in 1950. The loss of this traditional fishing ground forced vessels to operate in the coastal waters off Taiwan and subsequently resulted in the over-fishing of the Taiwan Strait.

To conserve the local marine ecosystem, fishing companies were required to upgrade their vessels to operate on more distant fishing grounds. The "Regulations Concerning the Importation of Motorized Fishing Vessels of Taiwan" [*Taiwansheng Guowai Dongli Yuchuan Shuru Guanli Banfa*] were issued at the beginning of 1951. These regulations had a number of significant effects. First, companies were again encouraged to increase vessel size. For example, pair-trawlers relocated to Taiwan now had to be at least eighty tons. Second, the regulations effectively forced additional modernization of fishing gear. All imported vessels were required to be equipped with a wireless and refrigeration equipment; the latter not only enabled the catch to be preserved longer but also enabled the vessels to operate on more remote grounds. Third, the new rules provided a legal basis for the introduction of foreign fishers and fishing techniques. During the colonial period, most new techniques came from Japan, but this was no longer possible after 1945. The timely implementation of the new regulations allowed foreign fishers and technicians to be hired as long as they did not exceed one-third of the crew. The measure reopened the door for the introduction of foreign fishing techniques.[18]

In the early postwar years the primary goal was to increase food production. The "Draft Plan Concerning Fisheries Production" [*Shueichan Shihye Zengchan Jihua Caoan*], proposed in 1951 by the Ministry of Economic

---

[17]Jhang, *Studies*, 27-28.

[18]*Ibid*.

Affairs, was an important case in point. At the time, Taiwan had a population of eight million people. If each inhabitant consumed fifty-two grams of meat per day, 152,000,000 kgs were required annually. But the livestock industry only produced 49,663,604 kilograms per year, while the catch of the fishing industry comprised 88,770,038 kilograms. Thus, there was a shortfall of 13,561,358 kgs. The government wanted to cover this through increased fisheries production.[19]

The development of fishing companies underwent two major changes in the early postwar years. First, the government became directly involved by establishing state-run fishing companies for the first time. Second, fishing enterprises that had been owned by tycoons in the colonial era disappeared and were replaced by a large number of small-scale private companies. To make this new system work, the government tried to ensure that the state-run companies were well-run and able to play a leading role. To accomplish this, the number of state-run companies was limited so that each would have sufficient funds. In contrast with the strict limit on the number of state-run companies, the introduction of numerous private firms was encouraged in the belief that competition would improve fishing techniques and prevent the rise of monopolies.[20]

The southward expansion of Taiwan's fisheries had been pioneered by the Japanese in the colonial era, and this thrust was reaffirmed by the postwar government, although the scope of its geographic considerations was different. The continental shelf off China between twenty-four and thirty degrees North latitude was traditionally viewed as the "northern fishing ground" by Taiwanese fishers. But because of the blockade by the Chinese communists, the northern grounds shrunk rapidly after 1949, forcing vessels to fish on the smaller grounds near Taiwan. To attempt to divert vessels to the South China Sea, Gulf of Tonkin and the seas off Hong Kong, the draft plan was amended to make Kaohsiung a more important supply port and base for pair-trawlers. Along with the existing vessels, the construction of thirty-one new units of pair-trawlers was planned, and it was estimated that the annual output of the pair-trawl fishery could reach as much as 28,000 tons. The waters off the coast of Southeast Asia between twenty degrees North and twenty degrees South were considered ideal for longline fishing. Longliners from Kaohsiung had been operating in the Bashi Strait, along the east and west coasts of the Philippines and the Sulu Sea, as well as in the Celebes Basin, since the colonial period. This earlier history of fishing in southern waters inspired the government to re-exploit these distant grounds. Fishing companies were encouraged to engage in longline fishing from Kaohsiung, and a large amount

---

[19] *Ibid.*, 303 and 306-307.

[20] *Ibid.*, 310-312.

of capital was allocated for the reconstruction of Kaohsiung Fishing Port, which made its facilities superior to those at Keelung. As a result, from the mid-1950s onward numerous trawl-fishing companies relocated to Kaohsiung to exploit the marine resources off Southeast Asia.[21]

During the seven-year Urgent Reconstruction Period, Taiwan's fisheries gradually recovered from the devastation of World War II. In 1952, output finally exceeded the catch record set in the colonial era. As a result, the Taiwanese fishery now bid entered another seven-year period which was referred to as the "Steady Recovery Period" ([*Yuye Hueifuci*]. Between 1953 and 1968 the fishing industry for the first time was integrated into the "4-Year Economic Construction Plans" [*Sihge Sihnian Jingjian Jihua*]. With continuous financial assistance from the government, the direction of Taiwan's fishing industry changed dramatically. The offshore fisheries expanded and matured, laying a solid foundation for the globally competitive fisheries of the future.[22]

The primary motive behind the "4-Year Plans" was to pursue the comprehensive development of the national economy. Various sectors, such as agriculture, manufacturing, mining and transport, were included, but fishing was still subsumed under agriculture. Nonetheless, the 4-Year Plans certainly benefited the fisheries. Large amounts of US aid were injected into the industry, and the private sector was encouraged to build larger vessels. In 1958, a more specific plan, the "Concentrative Utilization of Capital for the Exploitation of Marine Resources" [*Jijhong Yunyong Jhijin Kaifa Haiyang Jhiyuan*] was devised.[23] It called for the introduction of new fishing techniques and the upgrading of fisheries-related infrastructure, including markets, ice-manufacturing plants and ports. In addition, more overseas supply bases for distant-water longline fishing were to be established. The authorities expected that this assistance would attract more private investors to the industry.

By the end of the second four-year period in 1961, Taiwan's fishing industry had entered what is called the "Rapid Expansion Period" [*Yuye Kuojhangci*], which lasted until 1973. US aid, government subsidies and capital from the World Bank supported intensified efforts to expand the fisheries. As a consequence, by 1968 the aggregate tonnage of the distant-water pair- and single-trawlers had reached 55,249 tons, while longliners

---

[21]*Ibid.*, 312-313; Hu Singhua, *Maritime Taiwan* (Taipei, 2002, http://www.fa.gov.tw/chn/fish_culture/history/show_article.php?id=834, accessed 6 July 2005; Gu Tingfang, interview, 7 April 2002; and Lin Changren, interview, 17 May 2002.

[22]Sianchih, "Past, Present, and Future," 1.

[23]Huang Sianchih, "Economic Development and the Structural Adjustment of the Production of Taiwan's Fisheries" (unpublished paper presented to the Joint Annual Meeting of ROC Agricultural Groups, Taiwan, 1984), 2.

contributed another 85,300 tons. Further, fifty-three overseas supply bases for Taiwanese vessels had been established in the Pacific, Indian and Atlantic oceans, as well as the Mediterranean Sea. In addition to the rapid construction of modern vessels and supply bases, new fishing grounds were exploited. Comprehensive research had been carried out on tuna; on potential trawling grounds off Vietnam and the north coast of Borneo; and on coral and purse-seine grounds in the South China Sea. As well, the squid fishing grounds in the Bashi Strait and the demersal marine resources off the northern coast of Australia were also being examined. Although most of these new grounds were off Southeast Asia, Taiwanese fishers could now be found in virtually all of the world's major fishing zones.[24]

In 1968, when the 4-Year Plans ended, the Ministry of Economic Affairs and the FB decided to implement a "5-Year Programme for the Acceleration of Taiwan Fisheries Development" [*Jiasu Fajhan Taiwan Yuye Wunian Jihua*]. Unlike its predecessors, the new plan was directed especially at the fishing industry.[25] Under this new scheme, loans from the Asia Development Bank were allocated to companies for the construction of new vessels. Consequently, the fishing grounds exploited by Taiwanese fishers soon extended across the region and beyond. Those who had been fishing in the Bashi Strait now began to operate in the Banda Sea, and those who had been using the Celebes Sea now moved into the Indian and Pacific oceans.[26]

The continuous expansion of the Taiwanese fisheries encountered an unexpected challenge in 1973 when the energy crisis drove up fuel prices and seriously escalated the operating costs for Taiwanese fishing companies. Troubles for the fishing companies increased when some coastal nations proclaimed 200-mile Exclusive Economic Zones. Together, these factors ushered in an "Adjustment Period" [*Yuye Tiaojhengci*], which stretched from 1974 to 1987, during which the authorities were forced to reconsider the future direction of the fishery. Fisheries policies during this period became more conservative: licensing was restricted and completely suspended for small vessels. As well, trawler companies were no longer allowed to build new vessels if the tonnage did not match the requirements set by government.[27]

The fisheries policies of the first three decades after World War II can be summarized as follows. First, the emphasis was on the motorization of

---

[24]*Ibid.*; and *Development of the Fishing Industry*, 40-44.

[25]*Ibid.*, 11-12; and *Taiwan Fisheries* (Nantou, 1971), 17-18. The title "5-Year Programme for the Acceleration of Taiwan Fisheries Development" was used in *Taiwan Fisheries*, an official booklet.

[26]Sianchih, "Economic Development," 2.

[27]Sianchih, "Past, Present, and Future," 6.

vessels and the industrialization of the fisheries. These principles were first proposed in 1945 and pursued until 1973. The installation of modern gear and the increase in vessel size were always two of the key requirements for fishing companies wishing to import or build new vessels.

Second, the policies of the early postwar years encouraged the southward expansion of the fishing grounds. Even before the "Draft Plan Concerning Fisheries Production" was proposed in 1951, the authorities were already considering using Kaohsiung as a base for expansion into Southeast Asia. When the 4-Year Plans were implemented, numerous overseas supply bases for Taiwanese fishing vessels were established around the world, although the emphasis was always on Southeast Asia. We can look briefly at the activities of the Fisheries Research Institute (FRI) of Taiwan as an example of this thrust. The FRI's experimental vessel, *Haicing Hao*, was sent to the South China Sea to investigate demersal resources in 1954, 1960, 1961 and 1962. The fishing grounds investigated were gradually extended to the waters off Vietnam, Borneo, Malaysia and the Gulf of Thailand. From 1960 onwards, the government began to encourage trawling in these waters. After 1971, the FRI sent experimental vessels to explore the marine resources off northern Australia. From 1956 onwards, *Haicing Hao* also investigated tuna resources in the Indian and South Pacific oceans. Abundant stocks were discovered in the Banda Sea and off Sumatra and northwest Australia.[28]

Third, the postwar fisheries policies did not really deviate dramatically from the measures adopted by the Japanese colonial authorities. The prewar increase in vessel tonnage and the modernization of fishing equipment were always important priorities, and the exploitation of Southeast Asia's marine resources was considered a top priority, albeit for very different reasons. For the Japanese, large fishing vessels and modern equipment were of great help to its naval forces. Economic dominance of the southern fishing grounds also clearly assisted the Japanese empire's colonial expansion into Southeast Asia. By contrast, the postwar Taiwanese government wanted to modernize the fishing fleet to meet domestic food needs and to accelerate economic growth. The re-exploitation of Southeast Asia's marine resources was also about attempting to avoid possible problems stemming from tensions with China and to prevent over-fishing on the northern grounds.

Assistance from the International Community

In addition to guidance from the government, financial and technical assistance from the international community made a significant contribution to the reconstruction and development of Taiwan's fishing industry. The United Nations

---

[28]Jhang, *Studies*, 336; and *Development of the Fishing Industry*, 104-105. The mother-ship operating pattern was replaced by overseas supply bases.

Relief and Rehabilitation Administration (UNRRA) offered assistance in the form of materials and equipment after the end of World War II. In August 1948, a special meeting was convened by the "Rebuilding Committee" [*Shanhou Shihye Weiyuanhuei*] in Shanghai to discuss the allocation of these supplies. As a new territory of the ROC, Taiwan was allocated eight percent of the aid. In January 1949, most of the technical supplies were successfully shipped to Taiwan. The industrial supplies and machinery were estimated to be worth US $300,000, an enormous amount at the time. The Committee decided to sell some of this at market prices.[29]

Table 2.3
Uses of US Aid for Taiwan's Fishing Industry by the Council for US Aid, 1951-1959

Purpose	Currency*	Amount	Percent of Total Aid
Building New Vessels	$NT	107,161.00	62.0
	$US	1049.70	82.9
Purchase of Gear	$NT	1460.00	0.8
	$US	120.30	9.5
Repair of Vessels and Gear	$NT	13,016.00	7.5
Upgrading Fishing Ports and Harbours	$NT	10,471.00	6.1
Refrigeration Facilities	$NT	19,643.00	11.4
	$US	9.80	0.8
Aquaculture	$NT	11,133.00	6.4
	$US	7.30	0.6
Loans to Fishermen	$NT	4000.00	2.3
Research, Training and Management	$NT	5578.00	2.3
	$US	79.00	6.2
Other	$NT	249	0.2
Total	$NT	172,711.00	100.0
	$US	1266.10	100.0

Note: *US Aid was offered both in American dollars and New Taiwan dollars. In the 1950s, 1 US$ = 40 NT.

Source: *The Bluebook on US Aid and the Achievements of Taiwanese Fishing Industries* (Taipei, 1966), 10-11.

More than half the financial returns from this resale was used to build and upgrade fishing vessels – thirty new boats were built by fishermen's associations, and thirty-one diesel engines were installed. This went a long

[29] *Construction of the Agriculture and Forestry Industries*, 86-88.

way toward reviving the Taiwanese industry, even though most of the craft were small. In addition, fourteen new trucks were purchased by the fishermen's associations for the landward conveyance of fish. The employment of these vehicles helped to resuscitate the island-wide transportation system that had been devastated by the war.[30]

When the flow of supplies from UNRRA ended in 1951, the US began to provide massive financial support (see table 2.3) Two fisheries-related agencies received this aid. From 1951 through 1959, both the Council for US Aid and the JCRR administered the funds; after 1960, the task was consigned solely to the JCRR. This assistance and advice played a major role in the revival of Taiwan's fishing industry.[31]

Two key sources can tell us how US aid was used. One is the *Bluebook on US Aid and the Achievements of Taiwan's Fishing Industry* [*Taiwan Yuye Yunyong Meiyuan Chengguo Jiantao*], the official booklet on American aid to Taiwan's offshore fisheries; the other is the *Fisheries Tribune* [*Yuye Luntan*]. The former provides us with two tables that demonstrate how US aid was selectively used for rebuilding Taiwan's fishing industry. The first table (see table 2.3) depicts how "the Council for US Aid" used American capital to accelerate the development of Taiwan's fishing industry.

The majority of US funds received during this period was used to build new vessels; sixty-two percent of US aid provided in NT dollars and 82.9 percent of aid in US dollars was used for this purpose. From 1953 to 1958, 138 offshore fishing vessels and forty-six distant-water vessels were built with these grants.[32] It must be stressed, however, that American funds also played an important role in the overall development and construction of fisheries-related industries, such as upgrading fishing ports and harbours and establishing refrigeration facilities. The overall contribution of US aid during these years for Taiwan's fishing industry was both farsighted and comprehensive. Table 2.4 shows that the JCRR used these funds in the thirteen years between 1951 and 1964 in a similar fashion as the Council for US Aid had done in the previous decade. Nonetheless, there were some slight differences because most of its programmes were in the form of small grants aimed at the development of the coastal fisheries, including the purchase of small fishing boats and the building of minor fishing ports. It is important to note that despite the emphasis on small boats, nearly thirty percent of the funds were used to build distant-water fishing vessels.

---

[30]*Ibid.*

[31]*Bluebook*, 9-10; and *Fisheries Tribune*, No. 13 (1962), 7.

[32]*Bluebook*, 11.

Table 2.4
US Aid for the Development of Taiwan's Fishing Industry, 1951-1964, by the Sino-American Joint Commission on Rural Reconstruction

Purpose	Currency	Amount (unit: 1000)	Percent of Total
Building New Vessels and Engines	$NT	108,881	36.9*
Repair of Vessels and Gear	$NT	26,098	8.8
	$US	5	3.2
Refrigeration Facilities	$NT	11,010	3.7
Upgrading of Fishing Ports, Harbours and Moles	$NT	63,869	21.7
	$US	113	73.4
Research, Investigation and Training	$NT	7446	2.6
	$US	23	14.9
Loans to Fishermen	$NT	16,750	5.7
Production of Fishing Nets	$NT	2751	0.9
	$US	1	0.7
Aquaculture	$NT	28,404	9.6
	$US	11	7.1
Facilities in Fishing Communities	$NT	26,633	9
	$US	1	0.7
Others	$NT	3097	1.1
	$US	0	0
Total	$NT	294,939	100
	$US	154	100

Note: *NT $34,675,000, or nearly thirty percent of the total, was used to construct distant-water fishing vessels; the rest (NT $74,206,000) was used to develop the coastal fisheries.

Source: *Bluebook*, 12-13.

Taken together, the two tables show that a high proportion of financial assistance from the international community, and especially the US, was spent to build modern vessels. This distribution of funding demonstrates that the increase in vessel tonnage and the modernization of fishing techniques were consistently considered top priorities by the postwar fisheries authorities.

We may now consider how US aid was used specifically for the construction of fishing vessels using data from the *Fisheries Tribune*. The period (1951-1960) is slightly different than the time span in the previous tables, so it is understandable that the data are also slightly different. As table 2.5 shows, the largest expenditures in both US and NT dollars went to the distant-water sector. Compared to the coastal fisheries, expenditures on distant-water vessels in NT was twenty-seven times higher, while in US dollars it was twenty-five times greater. Moreover, the government spent thirty-two times more in NT on the construction of offshore fishing vessels than on coastal fisheries and twelve times more in US dollars. The amount of aid used to build coastal fishing vessels seems trivial by comparison. The data also remind us that the government concentrated most of its financial resources to build large distant-water and offshore fishing fleets. This pragmatic approach reflected the main development focus of its overall fisheries policy.

Table 2.5
US Aid for the Construction of Distant-Water Fishing Vessels, Offshore Fishing Vessels and the Coastal Fisheries, 1951-1960

		Amount	*	**	Tonnages
Construction of Distant-Water Fishing Vessels (including Large Diesel Engines)***	$NT	39,423,000	2700%	44	3710
	$US	949,900	2500%		
Construction of Offshore Fishing Vessels (including Diesel Engines and Accessories)****	$NT	46,375,700	3200%	177	3000
	$US	470,100	1200%		
Coastal fisheries (Purchase of Sampan Engines)	$NT	1,429,380	100%	NA	NA
	$US	38,000	100%		
Total	$NT	87,228,080			
	$US	1,458,000			

Notes: * denotes percentage compared to expenditures on coastal fisheries; ** denotes number of fishing vessels; *** most distant-water fishing vessels built with US aid were 100 and 150-ton longliners; **** most offshore fishing vessels built with US aid were twenty-five and thirty-ton craft.

Source: *Fisheries Tribune*, No. 13 (1962), 5.

The fiscal contributions of US aid to the fishing industry went even further, for they played a vital role in encouraging private investment.[33] Indeed, the stimulus provided led to more private investment, and the number of craft built with non-US funds was in fact greater during this period.[34]

In 1963, the penultimate year that Taiwan received US aid, the World Bank also granted loans to the fishing industry. The amount exceeded US $7.8 million and allowed the construction of sixteen tuna longliners, thirteen of which were 300-ton vessels, while the rest were 1000-ton ships. These vessels were expected to operate in the Atlantic Ocean. In 1966, the government negotiated with the World Bank for a second loan to build sixteen 250-ton tuna

---

[33] *Fisheries Tribune*, No. 11 (1961), 16.

[34] *Bluebook*, 14.

longliners and four 1500-ton trawlers. A boom in tuna prices on the international market, however, encouraged many more individuals to request loans to build 101 tuna longliners and eleven trawlers.[35] The overall tonnage of the vessels built with these new loans was much larger than the size of the fleets in the colonial period or the first decade of the postwar era since they were designed to operate in the Indian and Atlantic oceans rather than off Southeast Asia. It is clear that within two decades Taiwan's distant-water fishers had spread well beyond Southeast Asia to establish a global fishery.

Table 2.6
Growth of Distant-Water and Offshore Fisheries, 1945-1964

	Distant-Water fisheries		**Offshore Fisheries**	
	Annual Catch (tons)	**%**	**Annual Catch (tons)**	**%**
1945	68	NA	2600	NA
1952	18,514	100	29,696	100
1964	126,765	684*	161,151	542*

Note: * denotes percentage of the 1952 figure.

Sources: *The Bright Future of Taiwan's Fisheries* (Taipei, 1955), 7; and *Bluebook*, 41.

Table 2.7
Growth of Single- and Pair-Trawling and Longline Fishing, 1951-1964

	Single-Trawl Fishing		**Pair-Trawl Fishing**		**Distant-Water Longline Fishing**	
	Annual Catch (tons)	**%**	**Annual Catch (tons)**	**%**	**Annual Catch (tons)**	**%**
1951	7071	100	9259	100	NA	NA
1955	11,483	162*	23,772	257*	1118	100
1964	19,743	279*	97,495	1053*	9476	847**

Note: * denotes percentage of the 1951 figure. ** denotes percentage of the 1955 figure.

Source: *Bluebook*, 42.

Taiwan's postwar fishing industry made great strides in a relatively short time (see tables 2.6-2.8). If we use as a base total output in 1952 (the year Taiwan matched its prewar production record), the volume caught by

[35]*Fisheries Tribune*, No. 52 (1965), 3 and 5; and No. 58 (1966), 23.

distant-water fishers increased by a factor of 6.8 by 1964, the final year that Taiwan received US aid; over the same period, output from the offshore fisheries grew 5.4 times. When viewed by individual sector, the picture becomes even clearer. Compared to 1951, pair-trawling grew more than ten times by 1964, longline fishing expanded 8.4 times and single-trawling increased by 2.7 times. In less than two decades, Taiwan's fishing industry achieved unprecedented output in the offshore and distant-water sectors.

Table 2.8
Growth of Offshore Longline Fishing, 1953 and 1964

	Offshore Longline Fishing Annual Catch (tons)	Percent
1953	12,605	100
1964	15,944	126

Source: *Bluebook*, 43.

The development strategy for the postwar Taiwanese fishing industry was basically two-fold: to increase production using modern vessels and to build those ships with foreign financial assistance. These policies served as a domestic stimulus for the growth of Taiwan's fishing industry, while major financial assistance from UNRRA, the US and the World Bank served as key external catalysts. In the initial stage, the motive for developing the postwar fishing industry was to meet the urgent and growing demand for food for domestic consumption. But the combined incentives of government development policies and foreign financial aid led to an unprecedented growth, which in turn further encouraged government and private fishing companies to build a global industry. In the two decades after World War II, overseas supply bases were systematically established, and distant-water fishing vessels from Taiwan began to fish around the world. But these developments did not mean that the traditional distant-water grounds off Southeast Asia were abandoned. Detailed investigations into marine resources and potential fisheries were still being carried out there, and fisheries research vessels continued to explore waters as far away as off the coast of northern Australia.

Chapter 3

Kaohsiung Fishing Port and Its Fishing Ancillary Industries

The rapid growth of Taiwan's postwar fisheries resulted largely from rational national policies and international financial assistance. Nonetheless, there were also some local conditions that fostered its development, especially in Kaohsiung, although little attention has thus far been paid to them. Kaohsiung Fishing Port experienced dramatic changes over the course of the twentieth century. During the colonial era, it was simply a strategic forward base for Japan's southward-oriented fishing industry. But after World War II, it was gradually transformed into one of the most important regional centres for the new global fisheries. During this process, there was an urgent need for a modern shipbuilding industry in Kaohsiung. There was also a need for an efficient ice-manufacturing industry to cater to the fresh-fish trade.

The purpose of this chapter is to describe the construction of Kaohsiung Fishing Port, the development of its ancillary industries and how they contributed to the postwar revival of Taiwan's fisheries. I will begin by describing the building of Kaohsiung Fishing Port and its management. I will then focus on the ancillary industries, such as shipbuilding, ice manufacturing and fish processing, and assess their contribution to Kaohsiung's fishing industry. Finally, I will address southern Taiwan's fisheries education programme, considering especially how it supported fisheries development.

Reconstruction of Kaohsiung Fishing Port

During World War II, Takao Fishing Port was destroyed by American bombers. Most of its infrastructure was devastated; many vessels were sunk; and the waterways, channels and docks were obstructed by debris. This made the reconstruction and revitalization of the port a top priority for Taiwan's fisheries authorities.[1]

The postwar reconstruction of Kaohsiung Fishing Port resulted in tremendous changes in both its size and nature. Before 1945, the Takao (Kaohsiung) Fishing Port referred specifically to the Shōsento Fishing Harbour

---

[1] *A Brief Introduction to the Agricultural Industries, Vol. 10: Fisheries* (Taipei, 1962), 47.

[*Shaochuantou*], located in what is now called Gushan District. But in the postwar years Kaohsiung Fishing Port was greatly enlarged and renovated. In addition to rebuilding Gushan Fishing Port, a modern, distant-water fishing port built in 1967 in Cianjhen District attracted a large number of distant-water vessels originally based in Gushan. Several offshore harbours were also built in the Cijin District which encouraged a growing number of fishers to leave the coastal areas and fish off Southeast Asia. All these ports and harbours were generally considered to be part of Kaohsiung Fishing Port (see figure 3.1).[2]

Figure 3.1: Kaohsiung Fishing Port

Source: Courtesy of the author.

Among the various ports and harbours that comprised Kaohsiung Fishing Port, Gushan was the oldest. It had formerly been called the Shōsento Fishing Harbour in the colonial period. Shōsento comprised two harbours:

[2]*The Development of the Fishing Industry* (Nantou, 1971), 34. To distinguish the ports, the one located in Cianjhen District was called Cianjhen Fishing Port, while Shōsento Fishing Harbour was called Gushan Fishing Port.

Diyi Harbour, which was built in the colonial era, and Dier Harbour, which was constructed after the war. Before the completion of Cijin Fishing Port in 1967, most distant-water longliners and trawlers had used Gushan as their main supply base because of its well-developed infrastructure. The Gushan Fish Market was located near the harbour, and the catch could be auctioned as soon as it was unloaded. In turn, the market was connected to railways which enabled the fish to be conveyed rapidly to the major towns and cities on the island. Besides these two major advantages, the fishing vessels anchoring in Gushan also received a reliable supply of ice because most of Kaohsiung's icemakers were located there.[3]

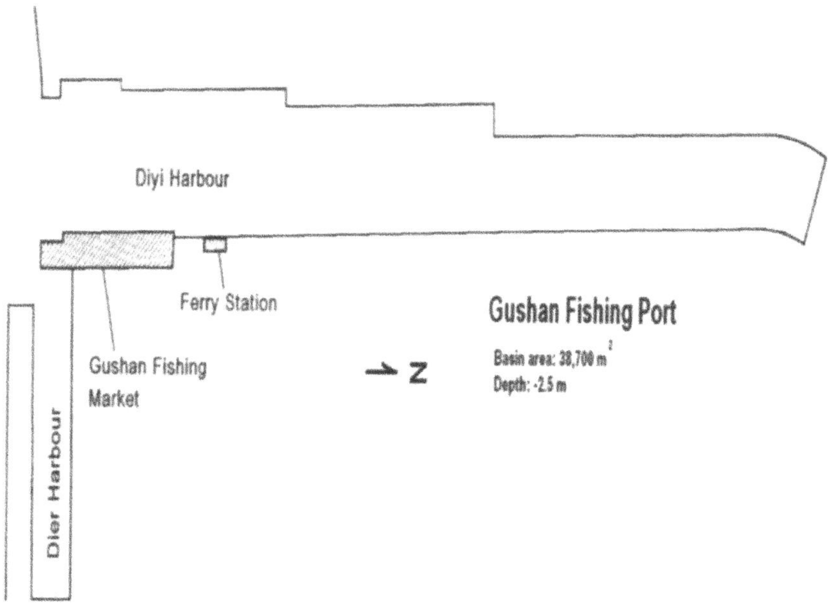

Figure 3.2: Gushan Fishing Port

Source: Courtesy of the author based on *Fishing Ports of Kaohsiung City* (Taipei, 2003), 9.

Figure 3.2 depicts the arrangement of the two harbours of Gushan Fishing Port, which was nestled into the town centre of Gushan District. The breakwater offered vessels a safe haven even in rough weather. Numerous offshore fishing vessels that fished in the Taiwan Strait and the South China

---

[3]Naitō Harukichi and Syu Jiwu, *The History of the Fishing Industry in Taiwan* (Taipei, 1957), 58-59.

Sea were attracted to the port, and many new jobs were created by the boom which lured large numbers of fishers from Penghu and Siao Liouciou to Gushan. With the steady growth of the distant-water longline and trawl fisheries, Gushan soon became congested. This was not rectified until 1967, when the new Cianjhen Fishing Port was completed. Thereafter, Gushan gradually became an auxiliary to Cianjhen; whenever Cianjhen was over-crowded, vessels were required to anchor at Gushan Fishing Port.[4]

Today Gushan Fishing Port is but a pale shadow of its former self. It became an old and messy anchorage after ferries and barges began to moor there on a regular basis. A large number of coastal and offshore fishing vessels also returned regularly during the Spring Dragon Boat Festival and the Mid-Autumn festival which caused the port to be even more congested at certain times. Moreover, problems caused by silting have also threatened operations.[5]

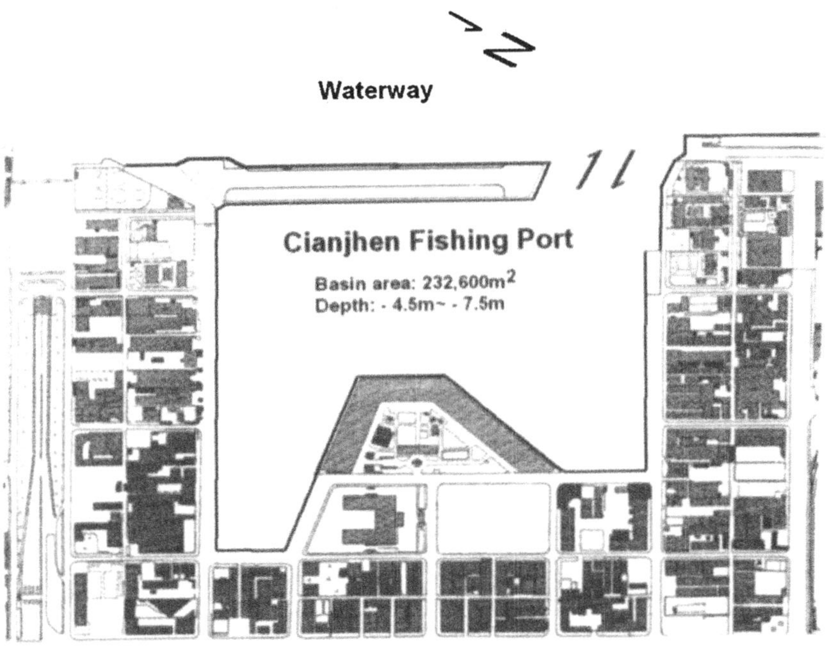

Figure 3.3: Cianjhen Fishing Port

Source: See figure 3.2.

[4]Hong Fucai, interview, 5 January 2003; Chen Shengli, interview, 21 May 2002; Zeng Jhengji, interview, 26 March 2002; and *Fishing Ports of Kaohsiung City* (Taipei, 2003), 9.

[5]*Fishing Ports of Kaohsiung City*, 9.

The construction of Cianjhen Fishing Port began in 1964 and was completed by 1967 (see figure 3.3). Its infrastructure, as befits Taiwan's largest distant-water fishing port, was more advanced and comprehensive than its competitors. There were more fuelling, water and ice stations than at Gushan. Ancillary industries, such as ice manufacturing, cold storage and fish processing, were clustered near the port. In addition, all the government's key fisheries-related agencies and institutes, such as the Weather Station's Kaohsiung Branch [*Cisiangjyu Gaosyong Cehoujhan*], Kaohsiung Training Centre ([*Gaosyong Yumin syunlian Jhongsin*], KTC), the Fisheries Research Institute Kaohsiung Branch [*Shueichan Shihyansuo Gaosyong Fensuo*] and the Taiwan Area Fishery Radio Station [*Taiwancyu Yuye Guangpo Diantai*] were located in Cianjhen. Blessed with these excellent resources, Cianjhen had a competitive advantage over Gushan, so it was not surprising that numerous fishing companies began to shift to there in the late 1960s.[6]

The establishment of Cianjhen Fishing Port was heralded as one of the most important milestones in the history of Taiwan's fishing industry. Fishing companies, ancillary industries and fisheries institutes were all assembled there and formed a self-sufficient environment for the growth of the distant-water fishery. The port not only contributed to the development and maintenance of Taiwan's fishery in Southeast Asia but also helped to establish a solid foundation for the growth of the global fisheries. Cianjhen can moor up to 370 distant-water fishing vessels. Although the port is huge, during peak seasons it is liable to be overcrowded, a problem which has often forced some vessels to anchor in Gushan instead.[7]

The Cijin District was formally a sandy coastal area with no harbours for offshore fishers; therefore, it was used only for the coastal fisheries. This began to change gradually due to the "secondary fishing ports" [*Erdenggang*] and "harbours" [*Chuanao*] built in the postwar years. Coastal fishers from the Cijin District, encouraged by the new facilities, began to fish in the Taiwan Strait and the South China Sea.[8]

The first port, Cihou Fishing Port, was built in 1952 to provide an anchorage for small fishing boats, most of which were engaged in the coastal and offshore fisheries. The second port, Shangjhuli Harbour, was also built for small vessels, but most of them operated in the Taiwan Strait, on the continental shelf near China and in the South China Sea. Meanwhile, the third

---

[6]*Ibid.*; *A Brief Introduction to the Fishing Industry of Taiwan* (Nantou, 1972), 32; *A Study of the Fishing Industry of Taiwan* (Taipei, 1974), 281; Chen, interview; Cai Wun'yu, interview, 29 May 2002; and Li Jijhao, interview, 22 March 2002.

[7]*Fishing Ports of Kaohsiung City*, 9.

[8]Cai Minghuei, interview, 5 July 2002; Cai Bian, interview, 3 July 2002; and Guo Shihfu, interview, 27 June 2002.

port, Jhongjhou Fishing Port, was constructed in 1952 on a sandy stretch of coast as a place where offshore vessels could anchor.[9]

In 1979, a major port construction programme was carried out in Cijin District; the government not only enlarged and renovated the three postwar fishing ports but also started to build Dashantou Harbour and Jhongcing Fishing Port. Because most ports and harbours, except Jhongcing, abut housing areas, there is little room today for expansion. Hence, during the typhoon season or major festivals, the ports all became overcrowded. Some fishing vessels were then forced to anchor at nearby shipyards. No matter how many new fishing ports the government built, they soon were unable to meet the demands of the growing number of vessels. This reflected the rapid progress of the offshore and distant-water fishing industries of Kaohsiung.[10]

Table 3.1
Financial Expenditures for the Construction of Fishing Ports in Taiwan, 1970
(NT $)

	From US Aid	From FB	From Local Govt.	Aggregate
Yilan County	7,206,584	11,990,480	6,437,534	25,634,598
Taipei County	6,645,537	13,356,846	6,626,493	26,628,876
Taoyuan County	6,916,067	5,689,746	7377640	19,983,453
Sinjhu County	1,014,200	4,090,900	2,319,561	7,424,661
Miaoli County	749,070	2,319,561	1,167,702	3,030,772
Taichung County		900,000	450,000	1,350,000
Changhua County		5,923,160	3,248,350	9,171,510
Yunlin County	1,301,700	5,573,752	2,495,000	9,370,452
Jiayi County	1,915,925	4,617,300	4,101,447	10,634,672
Tainan County	1,124,210	3,786,000	2,713,238	7,623,448
Kaohsiung County	3,207,450	9,145,020	4,972,357	17,324,827
Pingdong County	10,481,204	19,693,178	17,368,668	47,543,050
Taidong County	1,270,000	36,876,895	5,535,411	43,682,306
Hualian County		1,074,833	6,539,000	7,613,833
Penghu County	2,379,325	28,505,523	15,238,366	46,123,214
Keelung City	613,000	3,700,000	4,517,730	8,830,730
Tainan City		1,300,000	1,000,000	2,300,000
Kaohsiung City	15,784,000	25,464,000	20,825,100	62,073,100
Total	60,608,272	182,801,633	112,933,597	356,343,502

Source: *The Development of the Fishing Industry* (Nantou, 1971), 39-40.

---

[9]*Fishing Ports of Kaohsiung City*, 29; and *Kaohsiung Life Culture Education*, http://www.kh.edu.tw/, accessed 3 October 2004.

[10]*Fishing Ports of Kaohsiung City*, 37 and 43. When offshore fishing vessels came back to Kaohsiung for traditional festivals, the fishing ports became overcrowded.

Between 1945 and 1970, Kaohsiung Fishing Port received the greatest amount of construction subsidies from US aid, local government and the fisheries authorities; 17.4 percent of port construction expenditures were allocated to Kaohsiung, while its counterpart, Keelung Fishing Port, was given only 2.47 percent of the overall construction budget (see table 3.1). The exceptional position of Kaohsiung was clearly demonstrated by these comparative construction allocations.[11]

Ancillary Industries of Kaohsiung

Kaohsiung became Taiwan's largest industrial city during the colonial period. Major industries, such as sugar, fertilizer, aluminium, cement and papermaking were quite prosperous. Growth did not slow after 1945. The export of agricultural commodities, such as sugar, pineapples and bananas revived soon after the war. In 1956, modern facilities for transporting salt were built in the port, and in 1963 a warehouse for banana storage was constructed; it was converted into a refrigerated storehouse two years later. The canning industry also yielded exceptional results. Besides the revival of agricultural exports and traditional products, several modern industries had a crucial impact on its development. In 1957, the first plastics plant, Formosa Plastics Corporation [*Taisu Gongye Gongsih Gaosyongchang*], opened in Kaohsiung. By 1966, Taiwan's first "Export Processing Zone" [*Jiagong Chukoucyu*] was established. This attracted all sorts of modern industries to the city, which soon became the cradle of Taiwan's electronics industry.[12] The ancillary fishing industries were part of this urban development. Although not all of them made major contributions to the overall economy, all played important roles in supporting the revival of the fishing industry.

Most shipbuilding yards on Taiwan were badly damaged in the war, and at its conclusion the majority of Japanese shipbuilders and artisans were sent home. This explains why most of the fishing vessels built by local shipwrights in the early postwar years were small offshore and coastal boats. But since shipbuilding was an important prerequisite for a modern fishing industry, the reconstruction of shipbuilding capacity was thus a major goal for the fisheries authorities. Under the guidelines for "the motorization of fishing vessels and the industrialization of the fisheries" [*Yuye Gongyehua Yuchuan Donglihua*], shipyards in Kaohsiung, Keelung and Tainan were rebuilt

---

[11]*Development of the Fishing Industry*, 39-40. Table 3.1 shows that the government also invested large sums in building offshore fishing ports in other counties, such as Pingdong and Taidong, in order to keep fishing development balanced. See *The Bright Future of Taiwan's Fisheries* (Taipei, 1955), 25.

[12]Jheng Decing, *Looking Back on the History of Kaohsiung* (Kaohsiung, 2001), 144 and 205.

rapidly.[13] Kaohsiung soon became the centre of Taiwan's shipbuilding industry (see table 3.2).

Table 3.2
Shipbuilding in Taiwan, 1958

	Number of Shipyards	%	Total *Pings* of Shipyards	%
Kaohsiung Region	24	48	59,039	65.2
Keelung Region	8	16	9780	10.8
Tainan Region	6	12	4350	4.8
Penghu Region	3	6	4070	4.5
Pingdong Region	3	6	1460	1.6
Other	6	12	11,712	12.9
Aggregate	50	100	90,420	100.0

Note: A *ping* is equal to 3.3057 square meters.

Source: *Fisheries Tribune*, No. 16 (1962), 18.

Table 3.2 shows that in 1958 there were fifty shipyards on Taiwan (including the Penghu Archipelago), twenty-four of which were located in the Kaohsiung Region, accounting for fully forty-eight percent of Taiwan's shipbuilding capacity. Kaohsiung's dominance is even greater in terms of the physical area of the yards. Of a total of 90,420 *pings* across the island, the Kaohsiung Region accounted for 65.2 percent. Keelung, the centre of the distant-water fisheries in northern Taiwan, only accounted for 10.8 percent of the total area, less than one-sixth as much as in the Kaohsiung Region.

The shipbuilding industry of the Kaohsiung Region was flourishing by the end of the 1950s. Most of the fishing vessels sailing from Kaohsiung Fishing Port, including distant-water tuna longliners, single-trawlers and pair-trawlers, were designed and built by local shipbuilders. More important, these shipbuilders were capable of building larger craft than those constructed in Keelung. The vessels produced in Kaohsiung averaged between eighty and 200 tons, while fishing vessels built in Keelung were always less than 150 tons.[14] The shipbuilding industry in Kaohsiung was more efficient and had greater economies of scale than elsewhere in the country, which greatly benefited the development of its postwar fishing industry.

---

[13]*Fisheries Tribune*, No. 17 (1962), 15; and *Brief Introduction to the Fishing Industry*, 16.

[14]Lin Changren, interview, 17 May 2002; Gu Tingfang, interview, 7 April 2002; Pan San'guang, interview, 15 May 2002; and *Fisheries Tribune*, No. 16 (1962), 18.

Table 3.3
Daily Output of Ice-Manufacturing Plants in the Major Cities of Taiwan, 1951

City	Daily Output (tons)
Kaohsiung	262.0
Keelung	183.0
Taipei	177.5
Taichung	57.5
Tainan	56.0
Pingdong	55.0

Source: Naitō Harukichi and Syu Jiwu, *The History of the Fishing Industry in Taiwan* (Taipei, 1957), 58-59.

Ice played a significant role in keeping the catch fresh, allowed lengthy voyages and permitted longer periods of operation. The growth and prosperity of the ice-manufacturing industry was one of the most essential resources in an era when large refrigerators were still not widespread. The first ice-manufacturing plant in Taiwan was established in Taipei in 1901. Shortly afterward, the *Shinsho Seihyo* was established at Takao. The increasing demand for ice led to the continuous establishment of new ice-manufacturing plants around the island. To cater to the war effort and tighten controls over raw materials, in 1943 the entire ice-manufacturing industry was nationalized into the South Japan Fisheries-related Enterprise ([*Nanriben Yuye Tongjhih Hueishe*], SJFE), a government enterprise that controlled twenty ice-making factories which accounted for seventy-three percent of the island's output. Soon after the war, however, SJFE was taken over by the Fisheries Branch of the Agriculture and Forestry Company of Taiwan (AFCT). Subsequently, the ban on private ice manufacturers was lifted, and by September 1951, seven ice manufacturers were established in Kaohsiung: Kaohsiung Cold Storage 1st Plant [*Gaosyong Lengdongchang Diyichang*]; Kaohsiung Cold Storage 2nd Plant [*Gaosyong Lengdongchang Drerchang*]; Diyi Baoguanchu Ice Manufacturer [*Diyi Baoguanchu Jhihbingchang*]; Hechun Fisheries Ice Manufacturer [*Hechun Shueichan Jhihbingchang*]; Huanan Fisheries Ice Manufacturer [*Huanan Shueichan Jhihbingchang*]; Yurong Fisheries Ice Manufacturer [*Yurong Shueichan Jhihbingchang*]; and Sansiecheng Ice-manufacturer [*Sansiecheng Jhihbing-chang*]. The first three were owned and operated by the government, while the remainder were private. By 1951, the daily output of these seven was 262 tons, a total greater than any other city or county (see table 3.3).[15] Keelung City, for example, had a daily output of only 183 tons. The dominance of Kaohsiung's ice-manufacturing industry unquestionably helped to develop the distant-water fleets of southern Taiwan.

[15] Naitō and Syu, *History*, 57 and 59.

Most ice-manufacturers in Kaohsiung City were located near the port. Apart from one plant located in Cianjhen District and another in Ling'ya, the rest were located in Gushan, the nursery of Taiwan's distant-water fishing industry (see table 3.4).

Table 3.4
Main Ice Manufacturers in Kaohsiung City, 1951

	Output per Day (tons)	Location
Kaohsiung Cold Storage 1st Plant	140	Gushan
Kaohsiung Cold Storage 2nd Plant	50	Gushan
Diyi Baoguanchu Ice Manufacturer	30	Cianjhen
Hechun Fisheries Ice Manufacturer	20	Gushan
Huanan Fisheries Ice Manufacturer	15	Gushan
Yurong Fisheries Ice Manufacturer	4	Gushan
Sansiecheng Ice Manufacturer	3	Ling'ya
Total	262	

Source: See table 3.3.

Table 3.5
Development of the Ice-Manufacturing Industry in Taiwan, 1951-1964 (tons)

	Number of Manufacturers	%	Output	%	Ice for Fishing	%	Ice for Other Purposes
1951	22	100	109,605	100	78,181	100	33,557
1952	95		168,571		99,594		64,454
1953	105		183,581		107,780		68,754
1954	133		226,746		121,018		99,029
1959	203		392,177		223,258		103,395
1960	205		330,520		221,944		102,947
1961	238		397,306		256,417		135,378
1962	256		415,484		278,002		131,487
1963	273		456,167		311,987		139,909
1964	261	1186	472,261	431	314,832	403	148,442

Note: Data on ice manufacturing between 1955 and 1958 are incomplete.

Source: *The Bluebook on US Aid and the Achievements of the Taiwanese Fishing Industry* (Taipei, 1966), 53.

The ice-manufacturing industry grew rapidly in step with Taiwan's fishing industry. As table 3.5 demonstrates, between 1951 and 1964 (the last year Taiwan received US aid), the output of the ice-manufacturing industry grew four-fold. The industry's exponential growth was crucial to the development of distant-water fishing. As table 3.6 shows, by 1964 the port accounted for twenty-seven percent of the entire ice production in Taiwan, and nearly ninety percent of this was used by the fishing industry.

Table 3.6
Ice Manufacturing in the Major Cities and Counties of Taiwan, 1964

	Number of Manufac-turers	Output (tons)	%	Ice for Fish-ing (tons)	%	Other Purposes (tons)
Kaohsiung City	25	115,462	27.0	103,797	34.4	11,664
Keelung City	10	123,046	28.8	121,290	40.2	837
Taipei City	16	38,325	8.9		NA	38,325
Tainan County	16	24,715	5.8	3982	1.3	20,733
Tainan City	14	22,490	5.3	13,906	4.6	8583
Pingdong County	15	21,449	5.0	14,674	4.9	3518
Jiayi County	26	18,096	4.2	7024	2.3	7884
Kaohsiung County	16	16,658	3.9	9195	3	7463
Yilan County	13	15,737	3.7	14,182	4.7	1386
Taipei County	13	11,835	2.8	944	0.3	10,892
Penghu County	4	10,632	2.5	9529	3.2	973
Taichung County	17	9249	2.1	3439	1.1	4581
Total	185	427,694	100.0	301,952	100.0	116,839

Source: *Bluebook*, 54.

The development of the ice-manufacturing industry was so rapid in the two decades after 1945 that by the 1960s supply exceeded demand. During these years, refrigerators were still not common either in the ports or on board the vessels, but the plentiful supply of ice guaranteed Kaohsiung's fishing industry a sufficient supply throughout the 1950s and 1960s.[16]

Besides the development of the shipbuilding and ice-manufacturing industries, the fostering of comprehensive marketing channels for fish products also greatly benefited Kaohsiung's fishing sector. Significant information about the organization and development of the domestic marketing of fish products in the early postwar years can be found in an important book, *Research on the Prices and Marketing Channels of Fish Products in Taiwan* [*Taiwan Sianyu Yunsiao Yu Jiage Jhih Yanjiou*]. This nationwide marketing survey was conducted in 1955 by the Department of Agricultural Economics at National Taiwan University with support from the JCRR. Its express purpose was to

[16]*The Bluebook on US Aid and the Achievements of the Taiwanese Fishing Industry* (Taipei, 1966), 23-24.

describe and analyze the operation of fish markets and distribution channels for fish products.[17] As demonstrated in table 3.7, sales at the Kaohsiung Fish Market in 1955 accounted for 30.9 percent of the nation's volume by port fish markets.

Table 3.7
Sales Volumes of the Main Port Fish Markets in Taiwan, 1955 (tons)

	Sales Volume	%
Kaohsiung City	19,433	30.9
Suao	16,203	25.8
Keelung City	4,963	7.9
Tainan City	2,654	4.2
Donggang	2,359	3.7
Remainder	17,140	27.3
Total	62,752	100.0

Notes: Besides the five main port fish markets, there were thirty-eight others scattered along the coast of Taiwan.

Source: *Research on the Prices and Marketing Channels of Fish Products in Taiwan* (Taipei, 1956), 45.

The fish products unloaded at Kaohsiung Fish Market were not in the main consumed locally inhabitants. As table 3.8 shows, seventy percent of the fish sold at the Kaohsiung Fish Market were transported to other parts of Taiwan: twenty percent to Taipei; twenty percent to Taichung, Changhua and Jiayi; twenty percent to Tainan; and ten percent to the rest of southern Taiwan. The Kaohsiung Fish Market was the main supplier of fish products in Taiwan, and its marketing network was already well developed in the mid-1950s.

Table 3.8
Domestic Market Distribution of Fish Products Unloaded at Kaohsiung, 1954

	Fish Products (ton)	%
Taipei	3348	20
Taichung, Changhua, Jiayi	3348	20
Tainan	3348	20
Pingdong, Fongshan, Gangshan, Chaojho	1674	10
Kaohsiung	5022	30
	16,741	100

Source: *Research on the Prices and Marketing Channels*, 54.

---

[17]*Research on the Prices and Marketing Channels of Fish Products in Taiwan* (Taipei, 1956), 17.

In addition to the port marketing channels, a direct marketing system called the "collective marketing of distant-water boxed fish" [*Yuanyang Siangyu*] was adopted by individual fishers and distant-water companies in Keelung and Kaohsiung with the permission of the authorities.[18] This system avoided the inconvenience of onshore marketing and enabled the rapid distribution of fish products to inland markets. Fishing companies were permitted to transport their boxed products directly to inland markets as soon as they unloaded without having to deal directly with port fish markets.

Besides constructing an efficient nationwide marketing system, the authorities devoted considerable effort to developing fish-processing industries. Since the Taiwanese preferred fresh fish, and because the comprehensive nature of the marketing channels for fish products enabled this to be transported throughout Taiwan within a day, the fish-processing and canning industries were not very popular. The main processed fish products in the early postwar years comprised salted, dried and canned fish. Most was produced in small plants or workshops and often was sub-standard.[19]

The wide range of fish-processing industries in Kaohsiung provided the distant-water fishing companies with ample opportunities to sell all sorts of fish and marine products as raw material for processing. Seventeen kinds of major processed fish products are listed in table 3.9, and the fish-processing industry of Kaohsiung City produced seven of those types by 1953. Kaohsiung's fish-processing industry was twice as large as that of Keelung, the recognized centre of the distant-water fisheries in northern Taiwan.

In addition, technological breakthroughs in processing techniques helped Kaohsiung's fish-processing industries to prosper. A series of major improvements in processing techniques had been developed by the mid-1960s. In 1966, a method for vacuum packing dried fish was introduced with the assistance of the fisheries authorities, and it was followed shortly by the bonito-processing technique. From 1968 onward, the authorities put substantial effort into the research and development of canning and freezing techniques. As a result, fish unloaded in Kaohsiung, such as tuna, spearfish, bonito and pomfret, could be processed and preserved quickly. These technological achievements not only enabled the supply and demand of domestic markets to be balanced but also contributed to the increased exports of processed fish and fish products to international markets.[20]

---

[18]*Ibid.*, 66-68. Most of the catch by distant-water fishing vessels, especially trawlers, was packaged at sea. This is why particular products from trawlers were called "boxed fish" [*Siangyu*].

[19]*Bright Future of Taiwan's Fisheries*, 35; and *Development of the Fishing Industry*, 73-74.

[20]*Development of the Fishing Industry*, 74.

Table 3.9
Fish Processing Industries on Taiwan, 1953

		Keelung City	Kaohsiung City	Taipei City	Tainan Citv	Taichung City	Taipei Countv	Taichung County	Tainan County	Kaohsiung County	Yilan County	Penghu County	Changhua County	Jiayi County	Taidong County	Singjhu County	Pingdong County
Dried Fish	*Jhuganlei*						1				1	1					
Dried Fish	*Suganlei*																
Dried Fish	*Yanganlei*		1				1		1		1				1		1
Fish Balls	*Yuwan*				1	1			1				1	1			
Mullet Roe	*Wuyuzi*		1		1				1	1				1			
Shark's Fin	*Yuchih*	1	1		1					1	1					1	
Smoked Fish	*Syunjhihlei*						1								1		
Salted Fish	*Yancanglei*						1				1				1		
Canned Fish	*Guantoulei*		1							1	1		1				1
Other Edible							1										
Shell Powder	*Beikehuei*							1					1				
Fish Oil	*Yuyou*	1	1		1						1						
Cod Liver Oil	*Yuganyou*	1	1	1	1										1		
Fish Meat	*Yufen*										1						
Fish Scrap	*Yupo*		1								1						
	Jhuang-shihpin											1					
Others (Inedible)					1		1						1				
Total		3	7	1	6	1	6	1	3	3	8	2	4	2	4	1	2

Note: The first three categories in this table list types of dried fish that have slight differences in taste and presentation due to the manufacturing process.

Source: *The Bright Future of Taiwan's Fisheries* (Taipei, 1955), 36-37.

Fisheries Education in Kaohsiung and National Fisheries Research

The history of fishery education in Taiwan can be traced back to 1922 when two fishery vocational schools were established in Toko (Donggang) and Hōko (Penghu) by local governments. These pioneering developments were followed by several other fishery-related schools. None, however, was established in the Takao region in the colonial period.

The situation changed after World War II. The Keelung Fishery Vocational School [*Taiwan Zongdufu Shueichan Jiangsisuo*] and the Hōko Fishery Vocational School [*Magong Shueichan Busi Syuesiao*] were taken over by the government of the ROC in 1945 and renamed the Provincial Fishery

Vocational School of Keelung [*Shengli Jilong Shueichan Jhihye Syuesiao*] and the Provincial Fishery Vocational School of Penghu [*Shengli Penghu Shueichan Jhihye Syuesiao*], respectively. Meanwhile, a branch campus was set up in Kaohsiung by the Fishery Vocational School of Keelung. Its organizational structure, like its main campus in Keelung, consisted of three departments: fishing operations [*Yulaoke*], aquaculture [*Yangjhihke*] and fish processing [*Jhihzaoke*]. In the autumn of 1948, the Kaohsiung campus formally separated and became the Provincial Fishery Vocational School of Kaohsiung [*Shengli Gaosyong Shueichan Jhihye Syuesiao*]. Ten years later, the Department of Vessel Engineering [*Lunjike*] and Fisheries Management [*Jingyingke*] was established. In 1961 and 1962, the Anping and Donggang campuses were established by the Provincial Fishery Vocational School of Kaohsiung to keep pace with the development of the offshore and distant-water fisheries. In 1967, the Provincial Fishery Vocational School of Kaohsiung was officially upgraded to become the Provincial Kaohsiung Institute of Marine Technology [*Taiwan Shengli Gaosyong Haishih Jhuanke Syuesiao*], while its old Anping and Donggang campuses became independent. The former became the Provincial Fishery Vocational School of Tainan in 1967, and the latter became the Provincial Fishery Vocational School of Donggang the following year.[21]

Besides vocational education and training, the government of Taiwan also devoted considerable effort to the establishment of higher education in the fisheries. As early as 1953, an institute for the study of marine science and technology was established in Keelung; it consisted of seven fisheries-related departments and offered three-year courses to students in senior high schools. Ten years later, it was upgraded to a college offering full-time, four-year courses; in the process, it became the first maritime college [*Haiyang Cyueyuan*] in Taiwan. To develop experts in the field of marine biology, in 1967 a marine species unit [*Yuye Shengwuzu*] was established in the Department of Zoology [*Dongwusi*] at the National Taiwan University. The government also encouraged teachers in the fishery vocational schools to study overseas in order to be able to introduce advanced fishing techniques and to improve the quality of fishery education in Taiwan.[22]

Since the colonial era, the fisheries authorities have regarded experimental activities as important, and the role of marine research did not diminish in the immediate postwar decades. Two major fisheries research institutes were established in postwar Taiwan. One was the Marine Animals Research Institute [*Yuye Shengwu Yanjiousuo*] of the National Taiwan University, sponsored by the Ministry of Economic Affairs; the other was the

---

[21] *Brief Introduction to the Fishing Industry*, 53-54.

[22] *Ibid.*, 55.

Fisheries Research Institute of the Taiwan provincial government [*Taiwan Shengli Shueichan Shihyansuo*]. The former focused on the investigation of demersal species, while the latter concentrated on the improvement of fishing methods, the investigation of tuna resources, the exploitation of fishing grounds and aquaculture-related research. The research institute of the Taiwan provincial government also set up several branches in different areas of Taiwan, including one in Kaohsiung.[23] All these developments are summarized in figure 3.4.

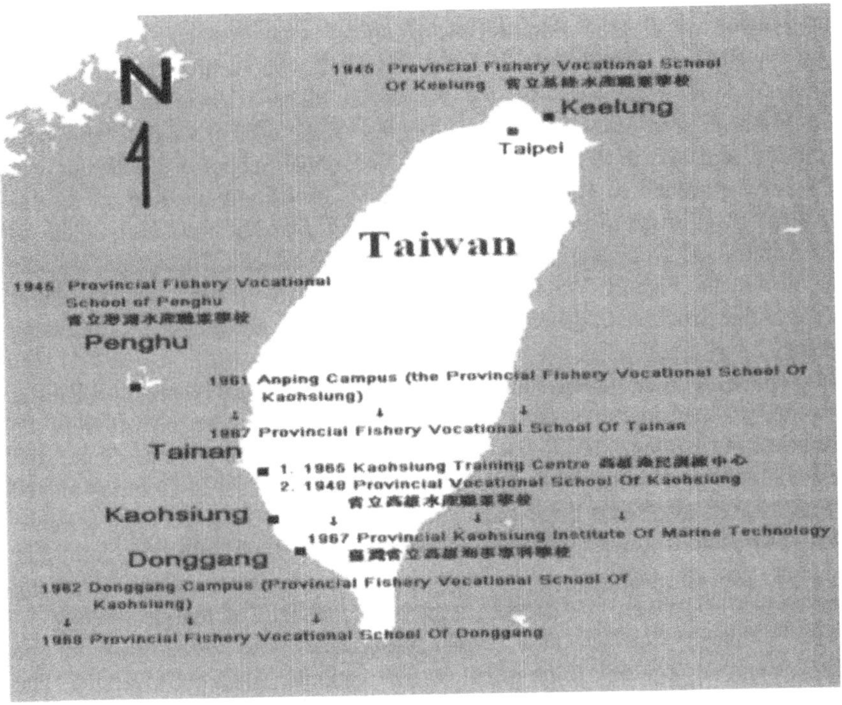

Figure 3.4: Fishery Education in Postwar Taiwan

Source: Courtesy of the author.

The diffusion of fisheries techniques in the colonial period depended greatly on the assistance of the Fishing Industries Association [*Shueishankai*] and Fishermen's Association [*Gyogyo Kumiai*]. Their fisheries-related workshops compensated to a certain extent for the lack of formal fisheries education. After World War II, however, the key role that the Fishermen's Associations played was not considered as important as before, and most

[23] *Ibid.*

workshops were sponsored directly by the fisheries authorities. Prior to 1964, there was no specific campus and no specially-assigned building for holding these classes. Instead, they were held intermittently in Kaohsiung, Keelung, Tainan, Penghu and other coastal areas. But in 1965, with subsidies from the JCRR, the Fisheries Bureau and the government of Kaohsiung City, a training centre for fishers, the Kaohsiung Training Centre ([*Gaosyong Yumin Syunlian Jhongsin*], KTC), was established. The KTC offered courses and programmes focussing on the latest fishing techniques and marine knowledge to engineering crews, radio officers and other staff on longliners and trawlers, and it also provided apprenticeship opportunities for deckhands. It was anticipated that 300-500 fishers could enrol in training programmes at the KTC every year. With professional fishing knowledge, most fishers graduating from the KTC would maintain a high level of performance and could hope to become officers on distant-water fishing vessels in four or five years time.[24]

Although fisheries-related academic institutes like the Fisheries Research Institute of the Taiwan provincial government and National Taiwan University are located in northern Taiwan, their research has supported the development of Kaohsiung's fishing industry. Furthermore, Donggang, Tainan and Penghu are situated close to Kaohsiung, and plentiful job opportunities in Kaohsiung lure most of the students from the nearby fishery vocational schools to work in the port after graduation.[25]

The construction of Kaohsiung Fishing Port served as a catalyst and base for the growth and development of Taiwan's fishing industry. In the early postwar years the reconstruction of Gushan Fishing Port encouraged Kaohsiung's longliner owners to exploit anew the marine resources of Southeast Asia. It also attracted Keelung's trawlers to relocate in Kaohsiung after the marine ecosystems of the northern fishing grounds declined in the mid-1950s. In a comparatively short period of time, fishing fleets from Kaohsiung began to dominate the fishing grounds of neighbouring Southeast Asia. Furthermore, the construction of Cianjhen Fishing Port in 1967 - one of the most advanced fishing ports in East Asia - represented a milestone in Taiwan's national maritime development strategy for the offshore and distant-water fisheries. A wide range of modern ancillary industries and efficient marketing systems also laid a solid foundation for the growth of these industrialized fisheries. Government agencies, like the weather station, fisheries research institute and marine communication centre, provided the latest fisheries-related information and technology to distant-water fishing companies. Plentiful job opportunities generated by the global fisheries also

[24]*Ibid.*, 16.

[25]Syu Yisin, interview, 5 May 2002.

attracted large numbers of fishers from Penghu and Siao Liouciou.[26] The presence of the KTC and nearby fishery vocational school also effectively improved the quality of the postwar labour force. The confluence of these excellent postwar conditions moulded Kaohsiung into one of the most modern and well-situated fishing ports in East Asia.

---

[26]See chapter 8.

Chapter 4

The Taiwanese Fishing Industry and the Military-Political Complex

Taiwan's postwar fishing industry was for a long time closely controlled and monitored by the government and the military. This level of state intervention directly affected the fishing companies and their activities at sea. Indeed, it was so exceptional that it deserves to be examined in some detail. In this chapter I focus on the relationships between the industry and the government, military and other politicians. Initially, I discuss the appointment of the heads of the fisheries authorities and show how the government kept the fishing industry under surveillance with the aid of the military. I then illustrate the nature of the interactions between the fishing industry, the military and the *Kuomintang* (KMT) government to show the kind of approaches the government and military used and the ways in which the fishing companies responded. In the last part of the chapter I will consider the issue of who could actually speak on behalf of the fishing industry given the level of government control. To provide part of the answer, I will examine the contributions of the parliamentarians towards the development the fishery. At the same time, I will discuss the strategies that the postwar industry in Kaohsiung used to increase its influence in governance and public affairs.

Military Control over the Fishing Authorities and the China Fishing Company (CFC)

Taiwan's fishing industry had close links with the military under the KMT regime. This unusual relationship was established for two important reasons. First, the offshore fishers were the only occupational group that could regularly leave Taiwan and contact people from China. Their mobility was considered a potential threat to Taiwan's political stability in the early postwar years. As a result, the government, especially the military intelligence branch, felt compelled to keep the industry under strict surveillance. Further, trawler fleets working off the Chinese coastal provinces needed naval escorts for safety. The government also believed that marine products would be a reliable food source for the military. In return, the fishing industry recognized that the military was an important and stable market for its products.

In order to monitor the fishing industry, high-ranking military officers were appointed to manage Taiwan's fisheries authorities. The most important of these in the early postwar years was the Taiwan Fisheries Production Committee ([*Taiwan Yuye Zengchan Weiyuanhuei*], or TFPC), which existed from 1951 until 1958. The TFPC comprised ten members representing various parties from both Taiwan and the US. Two came from the military; one from the Combined Logistics Command of the Taiwan Military, and the others from the US military consultative group.[1]

The appointment of the heads of the Fisheries Administration Agency ([*Yuye Guanlichu*], or FAA] between 1951 and October 1965) and the Fisheries Bureau ([*Yuyeju*], or FB) from 1965 until 1999 offers another example of the close ties between the military and the fishing industry. The first head of the FAA was Chen Baotai, a bureaucrat from China who was appointed mayor of Kaohsiung City in 1950. The position was turned over in 1952 to Liou Yongcyue, a high-ranking general who served as head of the FAA for a decade. In 1962, Liou Guosian took over. A former soldier, policeman and intelligence officer, Liou remained head of the FAA until his death in 1970. He was then succeeded temporarily by his deputy, Chen Banghao, and in late 1970 by Yao Daoyi, a former Rear Admiral. Yao had previously served as Deputy Director of the Telecommunications Department of the ROC Navy and as head of Naval Intelligence. More than simply a naval officer, he was one of the key figures in Taiwan's military intelligence community. The FAA/FB was in his charge for eleven and one-half years. After Yao retired in 1982, the position finally was given to a civil servant, Jhang Yitao.[2]

Two points need to be emphasized here. First, from 1951 to 1982 the fisheries authorities were controlled to a large extent by high-ranking military officers; their tenure illustrates the government's desire to keep the fishing

---

[1]Jhang Baoshu, *Studies in the Establishment of China's Fisheries* (Taipei, 1952), 337.

[2]"Online Directory of Government Officials," http://twinfo.ncl.edu.tw/tiqry/hypage.cgi?HYPAGE=search/search.sim.hpg&dtd_id=5, accessed 2 December 2004. Liou later held the posts of Chief of the Tainan County Police Bureau and Commissioner of the Taipei City Police Department. After these postings he returned to the military as Director of the Security and Control Department at the Taiwan Garrison Command. In 1965, the FAA was reorganized as the FB, but his position was not affected; see Fisheries Information Service, http://www.fa.gov.tw/organization/success/success.ph, accessed 2 December 2004. In the 1980s, the autocratic rule of the KMT regime began to moderate, which explains why the head of the FAA reverted to a civil official in 1982. In 1986, the first opposition party in postwar Taiwan, the Democratic Progressive Party, was established. In 1988, Dr. Lee Teng-hui came to power after more than four decades of rule by Chiang's regime. The lesser role of the military in the fishing industry resulted from these profound political changes.

industry under close surveillance. Second, both Liou Guosian and Yao Daoyi had intelligence and security backgrounds. This choice of personnel to lead the FAA/FB also suggests that the government wanted to use its intelligence service, especially naval intelligence, to control and regulate the fishing industry for almost three decades. Table 4.1 summarizes the backgrounds of the various heads of the FAA/FB from 1951 until its closure in 1986.

Table 4.1
Heads of the FAA/FB, 1951-1986

Name	Tenure in Office		Background
Chen Baotai	1951 Oct.-1952 July	8 months	Civil official
Liou Yongcyue	1952 July-1962 Dec.	10.5 years	Major General of the Army
Liou Guosian	1962 Dec,-1970 July	7.5 years	Major General of the Army
Chen Banghao	1970 July-1970 Dec.	0.5 years	Civil official
Yao Daoyi	1970 Dec.-1982 Jun.	11.5 years	Rear Admiral of the Navy
Jhang Yitao	1982 Jun.-1986 Dec.	4.5 years	Civil official

Source: Fisheries Information Service, http://www.fa.gov.tw/organization/success/success.ph, accessed 2 December 2004.

The military also maintained strong links with fishing companies. For example, the state-run fishing company, the China Fishing Company ([*Jhongguo Yuye Gongsih*]), was managed for a long period by two retired high-ranking officers, Chen Liang and Yuan Tongchou. After graduating from the Tokyo University of Agriculture, Chen returned to China to lecture at the Sixth Normal College of Jhejiang Province. But a teaching career did not suit him, so in 1925 he decided to join the army, where he rose quickly through the ranks to become a high-ranking officer in charge of military supplies and the provision of food. In 1946, his military career reached its apex when he was promoted to Deputy Commander-in-Chief of the Combined Logistics Command. Two years later, he left the military to work for the Ministry of Food Provision. Before becoming involved in Taiwan's fishing industry, he also briefly held several important government positions, including Mayor of Shanghai and Minister of Transportation and Communications.[3] Chen then became head of the Managing Agency of Fisheries Reconstruction Materials [*Yuye Shanhou Wuzih Guanlichu*], the predecessor of the CFC.

Chen's successor was Yuan Tongchou, a retired Lieutenant General. Yuan graduated from Nihon University in Japan and joined the army after his return to China. He was interested in the KMT's internal affairs and was heavily involved in the purge of communists within the government. In 1947,

---

[3]"Online Directory of Government Officials." See also Liou Shaotang (ed.), "The Profiles of the Republic's Celebrities, 236," *Biographical Literature*, LXV, No. 6 (1994), 133-134.

he was elected to the National Assembly and later worked as a trustee of the Taiwan Sugar Corporation. In 1963, he was appointed president of the CFC. Yuan's successor was Liou Guei, who had been a counsellor to the Ministry of Economic Affairs before being named president of the CFC in 1964. He was the first president without a military background, but the government's sudden decision to dismantle the CFC in 1965 also meant that he was the last.[4] Table 4.2 summarizes the background of the heads of the CFC from 1951 until its abolition in 1965.

Table 4.2
Presidents of the CFC, 1951-1965
(Including the Managing Agency of Fisheries Reconstruction Materials)

Name	**Tenure of Office**		**Background**
Chen Liang	1951-1963 Mar.	12 years	Deputy Commander-in-Chief
Yuan Tongchou	1963 Mar.-1964 July	1 year, 5 months	Lieutenant General of the Army
Liou Guei	1964 July-1965 Sep.	1 year, 1 month	Civil official

Source: *Fisheries Tribune*, No. 37 (1964), 4.

The CFC thus was effectively controlled by two retired high-ranking military officers for most of its existence (including the period of the Managing Agency of Fisheries Reconstruction Materials). The nature of their appointments suggests two things: first, that the government wanted to keep the fishing industry, especially the offshore and distant-water sectors, under strict surveillance; and second, that it wanted to boost food production, not only for the populace but also for the military through the control and development of Taiwan's fishing industry. This is why Chen Liang was trusted with the CFC presidency for more than ten years. Although the third president of the CFC was a civil servant, his term of service was brief.

Government Monitoring of the Fishing Communities

The military not only dominated the fishing authorities and state-run fishing companies but also monitored onshore fishing communities and fishing activities at sea. The archives of the Kaohsiung Fishing Boat Commercial Guild (KFBCG) provide detailed information about the nature of this. The KFBCG was officially established in 1965. Prior to 1965, the guild had existed

---

[4] "Online Directory of Government Officials" See also Liou Shaotang (ed.), "The Profiles of the Republic's Celebrities, 140," *Biographical Literature*, XLIX, No. 3 (1986), 147-148; and *Fisheries Tribune*, No. 22 (1963), 4; No. 37 (1964), 4; and No. 51 (1965), 3.

for several years as a sort of club, the Kaohsiung Pair-Trawler Owners Club ([*Kaohsiung Shoucaowang Lianyihuei*], or KPTOC), which was the oldest and largest fishing guild in Kaohsiung. The KFBCG periodically held a Member Representatives' Conference. Before this conference could be held, however, the KFBCG had to obtain permission from the municipal government. The conference applications were not only submitted to the Kaohsiung city government but also to the City Police Department, the local police post and the Kaohsiung Branch of the Taiwan Garrison Command [*Jingbei Zongbu*]. The KFBCG's archives show that representatives of the KMT, the military and the police always attended these meetings, although on most occasions they said nothing. It appears that their job was solely to monitor the conferences.[5]

The degree of social and political control exercised over the fishing community can also be seen through the standard procedures followed at these conferences. For example, at the second conference for the representatives of the fourth cohort in April 1972, all participants stood at attention and sang the national anthem at the opening. They then paid their respects to the national flag and Sun Chung-shan, the founder of the ROC, by bowing three times. Finally, the chairperson read the annual instructions of President Chiang Kai-shek to the conference. All such conferences had to start with these "quasi-religious" political rituals. Although the opening procedures of these conferences might seem extreme, in the years of the "White Terror," any KFBCG representative who attended had to follow them.[6]

Besides the conferences of KFBCG representatives, the leadership of Kaohsiung's fishing communities often met to discuss fisheries affairs at Guild Supervisors' Meetings or Combined Meetings for Guild Supervisors and Directors. The number of participants at these meetings was so small that the "quasi-religious political state ceremonies" were not practiced. Nor did representatives of the KMT, the military and the police attend. But the supervisors' meetings were still monitored by the "Guidance Division" [*Fudaozu*], a branch of the KFBCG comprising representatives from the military. Military representatives directly involved in Kaohsiung's fishing community date back to 1966 when a "Security Office" [*Ancyuanshih*] was established at request of the Taiwan Garrison Command. Unlike other civilian

---

[5]Kaohsiung Fishing Boat Commercial Guild (KFBCG), "2nd Conference Records of the 4th Cohort Member Representatives," 10 April 1972.

[6]*Ibid*. The term "White Terror" in Taiwanese history is used to describe the atmosphere of fear, threat and intimidation created by the KMT through surveillance, illegal imprisonment, assassination and the imposition of the death penalty under martial law from 1949 to 1987.

divisions of the KFBCG, the recruitment and dismissal of Security Office staff was handled by the Taiwan Garrison Command.[7]

The military not only monitored every onshore activity of the fishing community but also controlled activities at sea. After 1950, the military attempted to control fishing activities in the Taiwan Strait through the strict demarcation of the boundaries of the fishing grounds. According to a draft plan, the Strait was divided into two zones: the "Unrestricted Fishing Zone" [*Zihyou Buyucyu*] and the "Escorted Fishing Zone" [*Baohu Zuoyecyu*]. The latter was located in the northern part of the Strait, and only vessels with the proper equipment and the permission of the military were allowed to fish there. The complete withdrawal of the KMT's troops from China caused a dramatic reduction in the size of Taiwan's fishing grounds. After 1951, no trawlers from Taiwan, including those from Kaohsiung and Keelung, were permitted to fish in waters north of twenty-nine degrees North or within thirty nautical miles of China's coast. The government's decision to withdraw its troops from Dachen Island off the coast of China on 1 January 1954 further reduced the size of the grounds (see figure 4.1). Taiwanese fishing vessels were thereafter prohibited from operating in the waters north and west of the centre of the Taiwan Strait for military reasons.[8]

In November 1963 the government, ignoring the traditional rights of Taiwanese fishers, officially divided the Taiwan Strait into the "Unrestricted Fishing Zone" and the "Prohibited Zone" [*Jinjhih Buyucyu*].[9] This politically motivated measure created numerous problems for fishers, although in practice the government could not prevent them from operating in the Prohibited Zone. As well, many vessels accidentally entered this zone. In 1966, the government finally eased restrictions on the northern fishing grounds, allowing vessels to fish in there again - but only with the government's approval. In July 1973, the Prohibited Zone was renamed the "Conditional Fishing Zone" [*Shecing Buyucyu*] by the Ministry of Defence.[10]

But the restricted boundaries of the fishing grounds were not entirely due to the military tensions between Taiwan and China. The government's lack of trust in the fishers was another important reason. Fishers were one of the

---

[7]KFBCG, "21st Meeting Records of the 4th Cohort Directors," 1 December 1972; and "5th Meeting Records of the 6th Cohort Directors," 19 October 1976. The main purpose of the Taiwan Garrison Command was to maintain political stability. It was an intelligence branch which played a notorious role during the White Terror.

[8]*China Fisheries News*, 9 May 1960; and *A Study of the Fishing Industry of Taiwan* (Taipei, 1974), 34.

[9]KFBCG, "The Implementation Progress of the Work Plans of 1975," n.d.

[10]*Ibid.*; and *A Study of the Fishing Industry*, 34.

few groups who were able regularly to go overseas and who were therefore likely to encounter "red bandits" [*Gongfei*] - the Chinese communists. To prevent fishers from being directly affected by the political propaganda of the communists, the government and military tried to ensure the segregation of Taiwanese fishers from the Chinese people. Unfortunately, the geographical proximity of Taiwan and China made this extremely difficult.

Figure 4.1: Geographic Location of Dachen

Source: Courtesy of the author.

Apart from boundary maintenance, the military employed a number of other strategies to control Taiwanese fishing activities in the waters near the China coast. Initially, it attempted to indoctrinate Kaohsiung's fishers by holding anti-communist seminars. By January 1967, barely one year after the

Security Office was established, more than 1000 fishers in Kaohsiung had already attended seminars designed to provide elementary knowledge about intelligence gathering. As well, anti-communist bulletins were widely distributed to teach Kaohsiung fishers how to prevent "red bandits" from collecting sensitive information.[11] After 1973, the government also mandated that all vessels wishing to operate in the "Conditional Fishing Zone" had to do so in groups. Leading officers from each vessel were required to attend a four-hour workshop on national security. Any intelligence that a fishing vessel collected at sea had to be reported in detail to the Guidance Division. In order to ensure that these regulations were followed, the Taiwan Garrison Command assigned one escort for every twenty fishing vessels.[12]

In 1975 the military adopted an unusual method to further discourage vessels from working in Chinese waters by requiring all vessel owners to paint anti-communist slogans on the hulls of their vessels. These included "Strike down goddamned Mao Zedong!" [*Dadao Maofei Zedong*], and "Eliminate all red bandits!" [*Siaomie Gongchan Feidang*]. This extraordinary demand placed vessel owners in a dilemma. The KFBCG finally reached a compromise with the military to exempt vessels operating on distant fishing grounds, such as off northern Australia and New Zealand, from the requirement. In other words, only those vessels which operated in the Taiwan Strait and the East and South China seas were forced to paint slogans on their hulls.[13]

The military and the government also used a variety of methods to track the movement of fishers in distant waters. In the early postwar years, a shortage of civilian radio operators forced the fishing industry to rely heavily on retired military signalmen. With their assistance, fishing vessels, no matter where they were operating, were always under military surveillance. The use of military radio operators did not change until 1956 when the government began to train civilian radiomen. In that year, forty-eight fishers recommended by fishing companies were trained by the FAA for half a year. In 1958, forty additional persons were trained; twelve were from Keelung, thirty from

---

[11]KFBCG, "Progress Report, 1965 and 1966;" and "1st Conference Records of the 2nd Cohort Member Representatives," 20 January 1967.

[12]KFBCG, "9th Temporary Meeting Records of the 5th Cohort Directors," 5 December 1974; and *China Fisheries News*, 12 December 1960. The idea of working in groups had been suggested as early as December 1960 by the Provincial Fishermen's Association of Taiwan in order to assure safe operations. While the initial idea had nothing to do with national security, the Taiwan Garrison Command soon realized that asking fishing vessels to work in groups was impractical and soon postponed this plan. See *China Fisheries News*, 26 December 1960.

[13]KFBCG, "16th Combined Meeting Records of the 5th Cohort Supervisors and Directors," 28 May 1975.

Kaohsiung and fifteen were supplied by the "Veterans Affairs Commission" (VAC), one of the most important military organizations in postwar Taiwan. By 1958, only one out of three radio operators came from the military. On the other hand, from 1956 to 1976, twelve cohorts of radio operators were trained. The details, such as the number of trainees, their origins and where they were dispatched afterwards, changed only slightly for each group. But it was an unwritten law that all radiomen had to be members of the KMT. In 1964, the Taiwan Garrison Command took a further step to tighten security on board vessels by establishing the "Committee of Radiomen Deployment," which formally controlled the dispatch of all radiomen.[14]

With the radiomen's help, the government knew every move that Taiwanese fishers made at sea, and there are ample records in the KFBCG archives detailing the resulting rewards and punishments of fishers. Some received certificates for collecting intelligence for the military. On the other hand, some were punished for disobeying military instructions. Entry into prohibited zones, mooring in China's fishing ports or losing confidential documents, such as radio codebooks, could result in severe punishment. An example of such an incident was the capture of *Haipeng Nos. 3 and 5* by a Chinese patrol craft on 8 May 1960. In the course of the fracas, the fishing master and radioman failed to destroy the classified documents on board. Instead, they handed all of them, including the master plan, codebooks and safety routes of the *Dasia* Military Exercise, to the Chinese soldiers in exchange for their release and safe passage. This decision caused grave problems when they returned to Taiwan, where they were interrogated by the Taiwan Garrison Command and charged with revealing military secrets. They were consequently sentenced to prison for one year and four months, respectively.[15] This illustrates that the difficulties encountered by fishers operating in Chinese waters came not only from Chinese patrol craft but also from the military and the intelligence branch of the Taiwan government.

Military Mobilization and the Taiwanese Fishing Industry

The military not only kept the industry and fishing community under constant surveillance but also took measures to mobilize fishers to aid in military actions. In November 1959, the Ministry of National Defence ordered that all vessels over fifty tons had to participate in a week-long training exercise each year. Sometimes the exercise was carried out in Taiwanese zones, but

---

[14]Li Jijhao, interview, 22 March 2002; *China Fisheries News*, 24 October 1958; *Chinese Fisheries Weekly*, 28 June 1976; Jhong Ciourong, interview, 15 April 2002; and KPTOC, "6th Meeting Records of 1964," 9 June 1964.

[15]KFBCG, "4th Meeting Records of the 6th Cohort Directors," 19 August 1976; and *China Fisheries News*, 12 and 19 September 1960.

sometimes they were conducted in Chinese waters. In 1960, three trawlers from Kaohsiung, *Yongjin No. 3* and *Sinfeng Nos. 1 and 2*, were deployed on a twenty-five-day military mobilization called the *Jingteng* Exercise. Their task was to transport military supplies from Penghu to Jinmen, a small island near China, and then to return to Kaohsiung under heavy artillery fire from the Chinese coast. While not complicated, the mission was extremely dangerous.[16]

Although vessel owners received compensation from the government when their vessels were mobilized, most owners did not like these assignments for two reasons. First, the commandeering of fishing vessels paralyzed operations. Second, the hulls of fishing vessels were easily damaged during amphibious landings. In 1969, the companies heard rumours that the military might increase the amount of compensation for commandeered ships; without verifying the veracity of the information, they petitioned the government to take this step immediately.[17] Clearly, the companies were not happy with the remuneration they received for the damage or loss of their vessels.

When real engagements with China took place, there was no room for equivocation: fishing vessels were arbitrarily seized for military service. In August 1958, the "823 Bombardment," one of the largest and most severe engagements, took place. A number of fishing vessels were mobilized to transport military supplies and equipment between Taiwan and the outer islands. Their small size prevented them from being easily hit by artillery fire and, as a result, the use of fishing vessels to transport supplies proved an excellent strategy. Only one vessel, *Anhua No. 2*, was damaged by Chinese shells. Unfortunately, the fishing master, Song Yucheng, perished and one of the crew, Jhang Linsen, was wounded. The title of Navy Lieutenant was conferred upon Song posthumously, and compensation was paid to both families as well as to the vessel owner.[18]

Political Brainwashing and Efforts to Please the Government

Fishing newspapers and journals offer a unique perspective on the nature of the interaction between the industry and the government, especially since the

---

[16] *China Fisheries News*, 13 March 1958, 23 November 1959 and 4 July 1960.

[17] Gu Tingfang, interview, 1 July 2002; KFBCG, "13th Meeting Records of the 4th Cohort Directors," 28 February 1972; and *KFBCG Handbook 1969* (Kaohsiung, 1969).

[18] *China Fisheries News*, 2 March 1959. The requisition of fishing vessels for military purposes can frequently be found in countries with industrialized fishing fleets. During the Second World War, 816 English and Welsh trawlers were requisitioned by the military, and 200 steam drifters were taken into naval service; see Robb Robinson, *Trawling: The Rise and Fall of the British Trawl Fishery* (Exeter, 1998), 174.

government used them to disseminate its political ideology to fishers. In the early postwar years these publications were full of political commentary and anti-communist propaganda that was exaggerated and, in retrospect, often humorous. Basically, the stories from the fishing industry's newspapers can be classified into three different themes. First, some aimed at exaggerating the misery of Chinese fishers in order to highlight the decent standard of living of those in Taiwan. Second, some were designed to broadcast military policies and propaganda. Hence, the danger of fishing in Chinese waters was always exaggerated in these newspapers. Third, many stories emphasized how Chinese fishers hoped that Chiang Kai-shek would launch a counterattack and wipe out Beijing's communist regime. The story "15 Fishers Waved Farewell to Mazu and Returned to the Iron Curtain with Tears" [15 *Yumin Releiying-kuang Sibie Mazu Chongru Tiemu*] is a classic example.[19] The story described how a group of Chinese fishers fled to Mazu and enjoyed a free and happy life there for one week. But to prevent the persecution of their family members by the Communists, they finally returned to China. The Chinese fishers sincerely hoped that the "mighty and wise" President Chiang would liberate the Chinese people from slavery as soon as possible. Before they boarded their vessel for the return journey, all the fishers wept, stood solemnly before a portrait of Chiang and bowed as a sign of respect. The scenario was so touching that all present were moved to tears. In an era of high political tension and propaganda wars, all political information was strictly controlled by the government. The general public, especially those with limited education, may have genuinely believed many of these newspaper stories.

While the military and the government used newspapers for their own political purposes, the fishing industry also used them to publicize its loyalty to the KMT. To implore Chiang Kai-shek to continue his "reign" was also an astute way of pleasing the government. In 1959, when Chiang's presidency was drawing to a close, the Taiwan Fisheries Academy [*Taiwansheng Shueichan Syuehuei*], along with eight other research groups, proposed an "impromptu" motion to ask Chiang to continue. They not only dispatched telegrams to the Presidential Palace and the National Assembly but also asked *China Fisheries News* to publish their content on its front page.[20]

Besides flattering the KMT in the fishing newspapers and magazines, the fishing community also took specific public action to earn the government's trust. For example, when Chiang died in 1975, the KFBCG promised the government NT $100,000 for the construction of the Chiang Kai-shek Memorial Hall. In order to raise this money, the KFBCG levied a compulsory NT $500 tax on every unit of pair-trawlers and promised to donate

---

[19]*China Fisheries News*, 10 December 1956.

[20]*Ibid.*, 21 December 1959.

any outstanding balance itself. The following year, the Guild mobilized Kaohsiung's fishing community to commemorate the former leader at ceremonies held at Chiang's mausoleum. The KFBCG made it compulsory for each fishing company to send at least one representative to the ceremony.[21]

On the occasion of every traditional festival, such as the Spring Festival and the Mid-Autumn Festival, the fishing industry always gave the military money or gifts in recognition of its support. Similarly, when the 823 Bombardment was at its peak, the Kaohsiung fishing industry raised funds for national defence.[22] As well, the industry extended condolences to families who lost sons in battle. For example, during the "86 Naval Battle" in 1965, two Taiwanese warships - the 1250-ton *Jianmen* and the 300-ton *Jhangjiang* - were sunk by Chinese torpedo boats; thirty-two crewmen were captured and 197 killed in action. This battle was considered the most heroic instance of sea warfare in Taiwanese history. Kaohsiung's fishing industry was anxious to support the families that had directly suffered from this naval tragedy.[23]

The complex relationship between the government, military and fishing industry amounted to more than simple "control and be controlled." Occasionally, the military acted out of genuine support for the industry, especially for the vessel owners. For example, the military supported vessel owners when the fishers went on strike. Such strikes were viewed by the government as a serious threat to the fishery, as well as to national security and social stability. The strategy adopted by vessel owners to end these strikes was to seek help from both the Guidance Department and the military. A second point is that naval patrols on disputed fishing grounds often protected fishing vessels from being harassed by foreign patrol boats, although it appears that the assistance often was more psychological than military. Finally, the industry expected the military to purchase large amounts of fish products and to help to regulate domestic fish prices. In May 1974, the KFBCG for the first time asked the military to buy more fish. The KFBCG directors held trade talks with Marine Corps Headquarters that met with a positive response.[24]

---

[21]KFBCG, "16th Combined Meeting Records of the 5th Cohort Directors and Supervisors," 28 May 1975; and "4th Meeting Agenda of the 6th Cohort Directors," 19 August 1976.

[22]*China Fisheries News*, 12 September 1958.

[23]Chinese Warship Museum, http://vm.rdb.nthu.edu.tw/cwm/taiwan/4202.html, accessed 1 December 2004; and KFBCG, "3rd Combined Meeting Records of the 1st Cohort Directors and Supervisors," 21 August 1965.

[24]KFBCG, "13th Combined Meeting Records of the 5th Cohort Directors and Supervisors," 24 January 1975; and "9th Meeting Agenda of the 5th Cohort Directors," 11 May 1974.

Mainland Chinese Parliamentarians and Taiwan's Fishing Industry

Under such strict control and surveillance, who could speak on behalf of the interests of Kaohsiung's fishing industry when help was needed? According to the electoral system in the ROC, some members of parliament were elected as general members while others were elected as representatives of various trade organizations.[25] Ten National Assembly delegates and three legislators were elected from China's fishing industry in 1949. Strictly speaking, they were the political representatives of the Chinese fishers and fishing enterprises. After the Communists came to power, most of these delegates and legislators fled to Taiwan with the KMT. Under Chiang's regime they were allowed to renew their terms of office without facing re-election.

Table 4.3
National Assembly Delegates Elected from China's Fishing Industry, 1949

National Assembly Delegate	Political Affiliation
Dai Singti	KMT
Tang Chengzong	KMT
Wang Pingguei	KMT
Cyu Hancheng	KMT
Jhang Yuntai	KMT
Cai Yiren	KMT
Jhang Yunhan	KMT
Wu Zihliang	KMT
Jhang Zihjhu	Young China Party
Chen XX	Democratic Socialist Party

Note: At the time, journals, newspapers and books published in Taiwan were not allowed to name dignitaries who had surrendered to the Chinese communists. As a result, Chen's given name is unavailable.

Source: *Fisheries Tribune*, No. 98 (1972), 5.

Table 4.3 shows the National Assembly delegates elected by China's fishing industry in 1949. Not all of them, however, were deeply involved in the fishery. Jhang Zihjhu, for example, studied economics in France and lectured at various universities in China. As a key member of the Young China Party, he was enthusiastic about politics and ran several newspapers at different points in time. A career in education, politics and the mass media

[25]In the early postwar years, the legislative functions of the ROC were divided between the National Assembly and the Legislative *Yuan* (court). There was also a *Control Yuan*, which monitored the performance of the government.

occupied much of his time, and there is no evidence to indicate that he was ever involved in the fishing industry.[26]

Moreover, many of the early parliamentarians left the fishing industry after moving to Taiwan. Tang Chengzong had been general manager of the Shanghai Fish Market and worked for the China Mariners' Insurance Corporation [*Jhongguo Hanglian Chanwu Baosian Gongsih*] after moving to Taiwan. When he died of chronic rheumatism, the vacancy was filled by Lin Cifeng, who was from Kaohsiung County. Similarly, Jhang Yunhan worked as the Chief Secretary in the Agriculture and Forestry Company of Taiwan ([*Taiwan Nonglin Gongsih*], AFCT) and then left the fishing industry, working as Head of the Directors [*Jhuren Weiyuan*] at a KMT county branch. Wu Zihliang also devoted himself to the educational sector and left the fishing industry.[27]

A similar process occurred within the *Legislative Yuan* (the Legislative Parliament). Not all of the legislators came from the fishing industry or were heavily involved in fisheries (see table 4.4). Jiang Zuojhou joined the KMT shortly after graduation from university and spent most of his time in the army, ultimately becoming a high-ranking officer. He was elected to the *Legislative Yuan* as a representative of Shandong's fishing industry simply because he was head of the Shandong Fisheries Association [*Shandongsheng Yulian Huei*]. On the other hand, Sie Jhesheng received a Masters degree in Economics from Oxford University and had worked in both academic and financial circles in Guangdong Province. His election came about primarily because he participated in the establishment of the Hainan Fishing Company [*Hainan Shueichan Gongsih*].[28] How much these legislators knew about the fishing industry at the time of their election is questionable.

Table 4.4
Legislators Elected from China's Fishing Industry

Legislators	Political Affiliation
Jiang Zuojhou	KMT
Syu Panlong	
Sie Jhesheng	KMT

Source: See table 4.3.

---

[26]"Online Directory of Government Officials."

[27]*Fisheries Tribune*, No. 98 (1972), 5.

[28]"Online Directory of Government Officials."

Moreover, their comparatively weak links with the Chinese fishing industry deteriorated once they fled to Taiwan. Jiang Zuojhou suffered from chronic ill health and had only limited contact with the industry after moving to Taipei. In the *Legislative Yuan*, he never took the floor to advocate on behalf of the fishing industry. Syu Panlong formerly worked in Jhejiang's fishing industry. He relocated to Taiwan in 1948 but then mysteriously disappeared from public affairs for unknown reasons. Sie Jhesheng had worked for the preparatory committee of the Hainan Fishing Company in China. He stayed in Taiwan for only a brief period before leaving for Southeast Asia to promote Chinese culture and education.[29]

Although most of these legislators and National Assembly delegates proclaimed to be representatives of the fishing industry, few had much real contact with Taiwan's fishing circles. An exception was Dai Singti, who graduated from the Provincial Fishery School of Jhejiang [*Jhejiangsheng Shueichan Syuesiao*] and studied the manufacture of fishing gear at the Chiba Fisheries Research Institute in Japan [*Cianye Shueichan Shihyansuo*]. After returning to China, he worked at the Institute of Zoology and Botany [*Dongjhihwu Yanjiousuo*] of the Academia Sinica. Simultaneously, he held the post of Head of the Jhejiang Fisheries Association and was in charge of the training of local fishers. In 1946, he became the Principal of the National Jhapu Fishery School [*Jhapu Shueichan Syuesiao*]. The following year, he was elected to the National Assembly on behalf of the fishing industry. His enthusiasm for fisheries education did not wane after he moved to Taiwan. Dai established two maritime colleges and served as principal of three fisheries-related schools at different points in time. In 1949, he was appointed Principal of the Provincial Fishery Vocational School of Kaohsiung [*Shengli Kaohsiung Shueichan Jhihye Syuesiao*]. Four years later, he worked as the Director of the Steering Committee of the Provincial Marine College of Taiwan [*Taiwan Shengli Haishih Jhuanke Syuesiao*] and then served as Principal of this new college until 1956. In 1957, on the basis of his experience in fisheries education, he was elected as Head of the Taiwan Provincial Fishermen's Association. Then, in 1966, Dai once again pioneered the establishment of another college, the China Marine College [*Jhongguo Haishih Jhuanke Syuesiao*], where he served as Principal until he passed away in 1991. Dai maintained close contact with the industry, concerned himself directly with fisheries affairs and relentlessly promoted fisheries education.[30]

---

[29]*Fisheries Tribune*, No. 98 (1972), 5.

[30]Wu Tianren (ed.), *The Biographies of Those Who Engaged in the Fishing Industry of Early Taiwan* (Nantou, 1996), 16-19; *Fisheries Tribune*, No. 7 (1961), 7; and China College of Marine Technology and Commerce, http://www.ccmtc.edu.tw/cjcmt-web/cjcmt/secretary/eng-ver/pagej.htm, accessed 7 December 2004.

Besides those elected on behalf of the fishing industry, Jhang Baoshu, a legislator who was elected as a general member, also deserves to be mentioned. Jhang graduated from the Provincial Fishery College of Hebei [*Hebei Shengli Shueichan Jhuanke Syuesiao*] and studied the fisheries at Tokyo Imperial University. After returning to China, he held the posts of Head of the Hebei Provincial Fishermen's Association [*Hebeisheng Yuhuei Lishihjhang*], Secretary-General of the Hebei provincial government and leader of the KMT's Hebei branch. In 1948, he was elected as a legislator and entered the KMT political machine after moving to Taiwan. In addition to his political career, he did some teaching at the Provincial Marine College of Taiwan and the National Taiwan University. He also served as a committee member of the Taiwan Fisheries Production Committee. He published a semi-monthly fisheries magazine and wrote a number of books on the development of the fishing industry. Indeed, he was considered one of the most prolific writers in the field of the Taiwanese fisheries at that time, earning a PhD in Fishery Studies from the University of Tokyo in 1965. From 1968 to 1979, Jhang served as the Secretary-General of the KMT and became one of the most powerful political figures in Taiwan under Chiang.[31]

As highlighted above, only a few Chinese legislators and National Assembly delegates were genuinely interested in the development of Taiwan's fishing industry in the immediate postwar years. Three reasons help to explain this. First, they realized that their voters and electoral districts were not in Taiwan; hence, they were under no obligation to speak for Taiwan's fishing industry. Second, their permanent appointments in parliament made any effort to gain support from the industry unnecessary. Third, it is possible they felt a sense of cultural superiority and had little or no inclination to mingle with the local Taiwanese fishers. Thus, these politicians, with the exception of Dai Singti and Jhang Baoshu, kept their political and social distance from Taiwan's fishing industry.

In 1971, Taiwan was expelled from the United Nations. The loss of diplomatic support and recognition in the international arena raised the burning issue of the KMT's political legitimacy. With a view to preventing potential political unrest, the KMT government decided to absorb particular local elites into the national parliament. In 1972, fifty-three supplementary National Assembly delegates, fifty-one supplementary legislators and fifteen members of the *Control Yuan* were formally elected. According to the electoral law, one of the supplementary National Assembly seats and one in the legislature were

---

[31] *Fisheries Tribune*, No. 7 (1961), 7; *Sueichan Press*, http://www.taiwan-fisheries.com.tw/news/2002/news-021204-6.htm, accessed 5 December 2004; "Online Directory of Government Officials;" and *My E Gov*, http://www.gov.tw/EBOOKS/TWANNUAL/show_book.php?path=2_003_013, accessed 4 December 2004.

reserved for the fishing industry.[32] As a result, two young men from the local fishing industry, thirty-eight-year-old Huang Zecing and thirty-three-year-old Luo Chuanjin, were elected. It was the first time that the Taiwanese fishing industry had proper representatives in the national parliament.

Huang, who had graduated from Chung Hsing University, was supported by the Keelung fishing community. The prosperity of his family business made him a well-known figure in his home town. Luo Chuanjin also graduated from Chung Hsing University. In the postwar years his family moved to Kaohsiung and ran Hechun Fisheries Ice Manufacturer. The rapid development of Kaohsiung in the postwar years soon encouraged them to invest in the fishing industry. The enormous profits generated from their fishing-related investments enabled them to gain a foothold which led eventually to establishing their own fisheries enterprise.

These newly-elected representatives were required to represent the overall interests of the industry regardless of their electoral constituencies. Thus, Huang, although from Keelung, had to establish close political links with Kaohsiung's fishing circles. As a result, he attended almost every important meeting held within the Kaohsiung fishing industry. Similarly, Luo, from Kaohsiung, also had to pay attention to fisheries affairs in other parts of Taiwan. Luo did extremely well in both the fishing and political arena. In 1972, he was elected President of the Guild and was heavily involved in policy development in the Kaohsiung fishing industry. In 1983, he was elected to the *Legislative Yuan* by Taiwan's fishing industry and was continuously re-elected until his retirement in the late 1990s.[33]

The main points of this chapter about the relationships between the postwar Taiwan fishing industry and the military-political complex can be summarized as follows. In the first instance, the postwar fishing industry recognized that military surveillance and administrative intervention were inconvenient to their daily operations. Nonetheless, they complied with the military's instructions and seldom complained. This was partly because under martial law vessel owners had no choice but to cooperate with the military, and partly because they needed the military's help in providing security escorts and markets for their products. It should also be emphasized that the industry's attitude towards the government and military surveillance was affected by ethnic factors as well. Facing interference from the military and the KMT government, the mainland Chinese vessel owners who fled to Taiwan with the

---

[32]Wang Taisheng, "The Constitutional Law of New Taiwan: Derivation from Prewar Japan and Republican China," *National Taiwan University Law Journal*, XXXII, No. 1 (2003), 1-48; and *Fisheries Tribune*, No. 97 (1972), 2.

[33]Private School Promotion Foundation, http://www.schoolfund.org.tw/List/, accessed 19 December 2004.

KMT in 1949 were always the last to complain.[34] Furthermore, they frequently offered outright moral and material support to the military. Two factors make it possible to understand how the mainland Chinese vessel owners developed such a "complex." First, they had just escaped from the clutches of the Chinese communists in the late 1940s, and a pressing sense of political crisis and fear encouraged them to support the KMT government and the military. Second, under the rule of the KMT, the people who dominated the military were also from mainland China. It is possible that as a consequence of their similar ethnic backgrounds, the mainland Chinese vessel owners felt more comfortable dealing with these KMT groups.

The establishment of a strong relationship between the fishing industry and the political arena in Taiwan was a slow and painful process. Before the elections of supplementary National Assembly delegates, legislators and members of the *Control Yuan* were held in 1972, the parliaments and National Assembly were basically controlled by mainland Chinese. It was impossible at that time for Taiwan's fishing industry to send its own representatives to the national parliaments. The aspirations and needs of the fishing industry could therefore not be directly conveyed to the central government through the parliamentary system. Hence, the fishing industry adopted two approaches to compensate for this lack of national representation and recognition. First, they relied on help from mainland parliamentary members, even though few of them were inclined to speak on behalf of the fishing industry. Second, they sought attention and publicity from fisheries' magazines and newspapers in the hope that their appeals would attract the government's attention. Although they also wanted to get directly involved in politics, they did not make much progress until the elections of supplementary parliament members were held in 1972, at which time two representatives from Taiwan's fishing industry were finally elected. This political process took almost three decades to complete.

---

[34]All the fishing companies felt that the military requirements were burdensome, but their level of annoyance differed. While local vessel owners could not understand why the government did such "strange things," those from China always showed more understanding of the government's actions.

Chapter 5

The Kaohsiung Fishing Industry and Its Ancillary Industries

This chapter will examine the relationship between the Kaohsiung fishing industry and the ancillary industries. To make this manageable, I will limit my discussion to six of the most important ancillary industries: ice manufacturing, trucking, shipbuilding, equipment supply, security and stevedoring. The first three were well organized and financially independent of the fishing companies. When disputes arose, the government did not always side with the fishing industry. On the other hand, fishing equipment suppliers were small retail businesses or household operations which produced and sold rope, nets, cables and other fishing gear, while security companies and stevedoring firms comprised poor labourers who eked out a barely tolerable livelihood. Both the labourers and the small business owners had little market power or political clout and were not in as strong a bargaining position as the large shipyards, icemakers and transport companies. By examining the interactions between the fishing companies and these other industries, I will demonstrate the behavioural characteristics of fishing vessel owners and analyze the manner in which they dealt with threats from other industries with comparable social influence and financial strength, as well as people from other social strata like the equipment suppliers, stevedores and security companies.

The "War" between the Vessel Owners and Ice Manufacturers

The first group with which the fishing industry fought fiercely in the early postwar era was the ice manufacturers. In the 1950s, refrigeration equipment was not yet widely used on fishing vessels, so having a reliable and reasonably priced ice supply was critically important. Fortunately, neither the price nor supply of ice were major issues. The situation changed dramatically, however, when ice manufacturers in Nanfang'ao unexpectedly increased prices without governmental authorization in May 1958.

The rise in ice prices at Nanfang'ao, a fishing port in the northeastern corner of Taiwan (see figure 5.1), triggered nationwide price fluctuations. The Ice Manufacturers Association of Taiwan Province ([*Taiwansheng Jhihbingye Gonghuei*], or IMATP) took similar concerted action and encouraged its members (except those in Taipei) to follow Nanfang'ao's example after 10 May. As a result, the price of ice for the fishing industry soared from NT $15.8 to NT $18 for 300 pounds; NT $17.4 to NT $19.8 for 330 pounds; and NT $21 to NT $24 for 400 pounds. In short, ice prices increased on average by about fourteen percent. IMATP claimed that the price hike reflected increased costs and adjustments in the exchange rate. Furthermore, supply and demand were no longer in balance, and ice prices had fallen far below their "real market value." The rise was designed to make prices return to that "real value."[1]

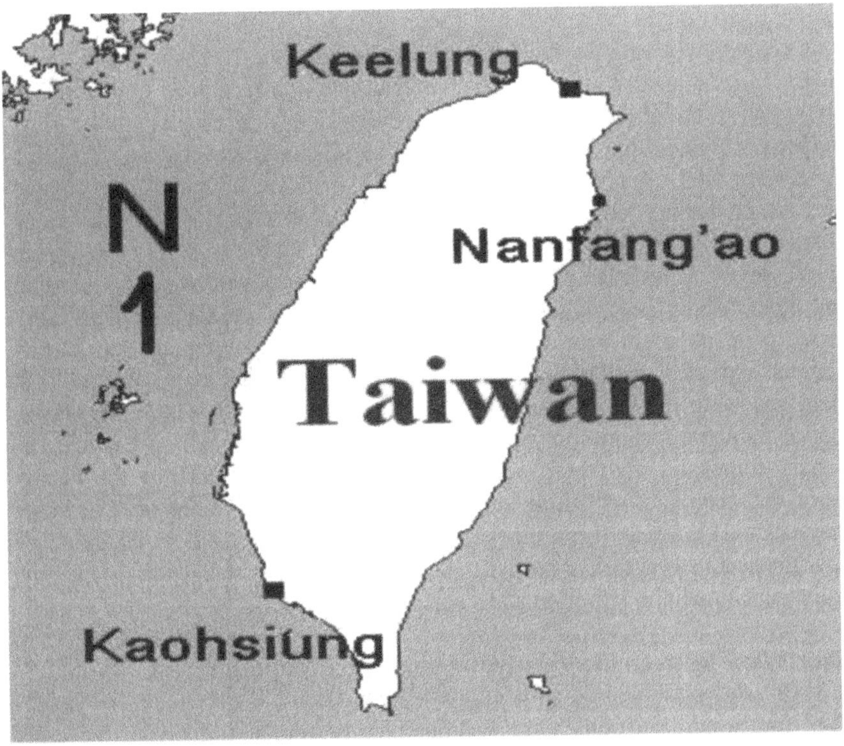

Figure 5.1: Location of Nanfang'ao

Source: Courtesy of the author.

In fact, IMATP had a very good reason to raise prices. The average exchange rate between the New Taiwan Dollar and the US Dollar had fallen from NT $10.30 in 1951 to NT $24.78 in 1958. To recover their financial losses due to this currency fluctuation, ice manufacturers had to demand more money from their customers. Nonetheless, the price rise caused real concern in Taiwan's fishing industry, which mounted a strong opposition. The fishing

[1]*China Fisheries News*, 22 and 29 May 1958.

industry argued that since water for ice manufacture was not imported, the increase in prices could not be blamed on the exchange rate. Moreover, the rise in power rates (which was real) did not significantly affect the operating costs of the ice manufacturers because it accounted for only a small portion of total costs. Finally, they argued, the rise in ice prices would discourage both the fishing industry and general public from purchasing ice, which would further worsen the problem of overproduction. The vessel owners were not persuaded by the manufacturers' rhetoric but instead were "very angry" with the prise rise.[2]

The Fisheries Administration Agency (FAA) was placed under significant pressure by both the Fishermen's Association and the Ministry of Economic Affairs. Vessel owners believed that the ice manufacturers were colluding to manipulate prices. Initially, the Construction Bureau [*Jiansheting*] of the provincial government sided with the fishing industry and issued an administrative decree on 22 July 1958 prohibiting any further price rises. It did this for two reasons. First, according to the Management Regulations of Agriculture, Mining, Manufacturing and Commerce in a State of National Emergency [*Feichang Shihci Nongkuanggongshang Guanli Tiaoli*], ice production was under the government's direct control; hence ice manufacturers were not supposed to raise prices without the prior approval of the government. Second, from an economic perspective, it judged that the price was already favourable for the manufacturers.[3]

As a consequence of this government intervention, ice manufacturers were forced to lower prices from 1 August. Unfortunately, this reduction proved to be merely a temporary victory for the fishing industry because on 7 September the Construction Bureau changed its position and allowed the manufacturers to raise ice prices. Vessel owners felt betrayed when they heard the news. From their perspective, ice manufacturing was an extremely lucrative business. Even before prices rose, the returns from the business averaged 130 percent; after the price rise, profits jumped to 162 percent, which the fishing companies considered exorbitant. There were 187 ice manufacturers in Taiwan in 1958, and their daily output had reached 2009 tons. Yet only thirty-nine percent of the ice produced was actually sold. Supply had outstripped demand, and hence ice prices should have fallen. Moreover, the average price of fish had declined for several years; if the ice manufacturers insisted on raising prices at this crucial moment, the fishing

---

[2]Paul C.H. Chiu, "Money and Financial Markets: the Domestic Perspective," in Gustav Ranis (ed.), *Taiwan: From Developing to Mature Economy* (Boulder, CO, 1992), 123.

[3]*China Fisheries News*, 3 September 1956 and 26 September and 24 October 1958.

companies argued, Taiwan's fishing industry might collapse.[4] As well, Taiwan and China were still at war, and any major fluctuation in the price of commodities and staples like ice and fish could well panic the general public.[5]

Representatives of the vessel owners and ice manufacturers continued to lobby the authorities, and in March 1959 two decisions were finally made. First, with a view to avoiding monopoly, ice manufacturers were prohibited from colluding. Second, and most important, the fluctuation in the price of ice was to be determined by the market - government would no longer intervene.[6]

By mid-1959, with the market now deciding prices, fierce competition ensued between ice manufacturers in Kaohsiung. Prices slumped from NT $15.8 to NT $12 per 300-pound unit. To try to counteract this, Hu Jidai, the head of IMATP, encouraged ice manufacturers in Kaohsiung to collude. Hu collaborated with the Hechun Fisheries Ice Manufacturer [*Hechun Shueichan Jhihbingchang*], the largest private producer in Kaohsiung, by appointing its owner, Luo Dengzeng, as the director of IMATP's Kaohsiung branch. The Kaohsiung Cold Storage 1st and 2nd Plants [*Kaohsiung Lengdongchang*], after obtaining permission from their parent company, the Fisheries Branch of the Agriculture and Forestry Company of Taiwan (AFCT), also joined this collusive scheme.[7]

The output of the Kaohsiung Cold Storage 1st and 2nd Plants and the Hechun Fisheries Ice Manufacturer accounted for a major portion of Kaohsiung's ice production in the 1950s.[8] As a result, prices in Kaohsiung rose again on 21 September, generating another wave of confrontations between vessel owners and ice manufacturers. The conflict came to a head a year later when ice manufacturers took concerted action to crack down on the Yuandong Fishing Company for failing to pay its debts to the Kaohsiung Cold Storage 1st and 2nd Plants on time. While his dispute was of little concern to outsiders,

---

[4]*Ibid.*, 26 September and 24 October 1958.

[5]*Ibid.*, 26 September 1958; and Hsueh Hua-yuen, "The 823 Artillery War and Is Historical Significance," *Taiwan History Association*, http://www.twhistory.org.tw/20010820.htm, accessed 4 February 2005. The 823 Artillery War [*823 Paojhan*] began on 23 August 1958 and continued to 7 January 1959. It was the largest and most severe battle between Taiwan and China.

[6]*China Fisheries News*, 24 October 1958 and 2 March 1959. The FAA promised a quick response after analyzing the relevant materials. The Construction Bureau demanded that the Ministry of Economic Affairs provide further instructions.

[7]*Ibid.*, 26 October 1959. An important point to emphasize is that business collusion was prohibited.

[8]See chapter 3. The aggregate output of Kaohsiung's ice manufacturers was 262 tons per day. Three of them accounted for 80.1 percent of the total in 1951.

the heavy-handed and coordinated action of the ice manufacturers suggested that allegations of collusion were correct. The vessel owners did not take long to file strong complaints with the authorities.[9]

The Kaohsiung city government tried to mediate the dispute and on 15 September 1960 made several proposals. One was that the Kaohsiung Fishermen's Association cancel its action against the ice manufacturers, and that in return the manufacturers agree to sell ice to vessel owners once again. In addition, changes in ice prices should be decided by the provincial government. Ice manufacturers, not surprisingly, did not approve of the city's suggestions, and the negotiations were aborted. The ice manufacturers proclaimed the necessity of a single price for ice to prevent the reoccurrence of "malicious competition" resulting from overproduction; hence, in their view, any price increase should not be misinterpreted as collusion. Besides, they alleged, the production cost of ice was NT $23.4 per unit, but the actual selling price was only NT $22. Not surprisingly, the manufacturers claimed large losses and insisted on raising prices.[10]

From the standpoint of the vessel owners, ice prices should not have been raised without governmental permission, or at least without first informing the fishing industry. As for the accusation of collusion, vessel owners had substantial evidence. First, a "bag man" was employed by Kaohsiung's ice manufacturers; that is, no matter from which manufacturer owners purchased ice, they had to pay the same person. Second, the money was deposited in a joint account at the Third Credit Cooperative [*Disan Sinyong Hezuoshe*] under the names of Luo Dengzeng, Sha Dasin and Hong Jieshan, three leading figures in the Kaohsiung ice manufacturing industry. Vessel owners also revealed that the money was then returned to the "real ice providers" at the cut-rate price of NT $6 per unit, with the rest distributed to the ice manufacturers who joined the collusion scheme (including the real providers) according to the size of their business. Vessel owners urged the authorities to audit the ice manufacturers' bank accounts. They further charged that some of the manufacturers did not sell any ice, yet money was still deposited in their accounts. These funds were the profits derived from collusion, and the accounts would verify this.[11]

With this strong evidence in hand, fishing vessel owners successfully convinced the relevant authorities that collusion was occurring. The provincial government asked the Kaohsiung city government to ban this practice, but the

---

[9] *China Fisheries News*, 26 October 1959 and 26 September 1960. None of the ice manufacturers was willing to sell ice to Yuandong, even though it promised to pay in cash. This was viewed as a warning sign by Kaohsiung's fishing industry.

[10] *Ibid.*, 19 and 26 September 1960.

[11] *Ibid.*, 26 September and 14 November 1960.

Kaohsiung City Police Department not only refused to execute this order but also questioned its legitimacy. It claimed that the Ministry of Economic Affairs had stated publicly that government would not intervene in the dispute over ice prices. The police emphasized that they were not partial to the ice manufacturers; they had spent much effort investigating the case, had carefully interrogated every ice manufacturer in Kaohsiung and claimed not to have found any solid evidence of collusion. Hence, they were not prepared to take any action against the manufacturers. Kaohsiung's vessel owners were extremely upset by this and urged the government to investigate whether the police had accepted bribes from the ice manufacturers.[12]

The situation ended in a deadlock that the authorities could not think of a way to break. To solve the problem, the vessel owners in 1961 announced plans to establish their own ice-manufacturing plant, the United Ice Manufacturing Plant for the Fishing Industries [*Yuye Lianhe Jhihbingchang*]. This plan, however, never came to fruition because the AFCT's Kaohsiung Cold Storage 1st and 2nd Plants, bowing to pressure from the FAA, decided to cooperate to break the ice-manufacturers' ring.[13]

Under the new arrangement, vessels owner were to pay an advance of NT$3 million to the AFCT, two-thirds as a deposit and the rest as insurance against any possible default. Each 300-pound block of ice would cost NT $18, and any price change would have to be approved by the FAA. The AFCT would guarantee 3000 tons of ice per month to vessel owners, one-third of which was allocated to longliners and two-thirds to trawlers. If consumption fell below 3000 tons, the vessel owners would have to cover AFCT's financial losses. It was estimated that through this cooperative scheme each unit of pair-trawlers would save more than NT $30,000 in operating costs per year. To avoid possible harassment from ice manufacturers, the negotiations were conducted behind closed doors.[14]

The cooperative plan formally came into effect on 15 April 1962. To handle its implementation, the vessel owners established the Ice Supply

---

[12]*Ibid.*, 12 December 1960.

[13]*Ibid.*, 26 December 1960 and 13 February 1961; and Kaohsiung Pair-Trawler Owners Club (KPTOC), "1st Meeting Records of the KPTOC in 1962," 24 February 1962. The plan to set up an ice-manufacturing plant miscarried because its construction proved prohibitively expensive.

[14]KPTOC, "1st Meeting Records of the KPTOC in 1962." According to the Kaohsiung Fishing Boat Commercial Guild (KFBCG), "The Combined Temporary Meeting Records of the 1st Cohort Directors and Supervisors," 29 May 1965, ice prices increased by ten percent in May 1965 with the consent of vessel owners, contracted ice manufacturers and the FAA. This suggests that this cooperative system worked well.

Committee [*Yubing Gongying Siaozu*], which comprised seven members, two from the longline fishing sector and five from the trawling industry. Since the amount of ice supplied by the Kaohsiung Cold Storage 1st and 2nd Plants could not always meet the fishing companies' needs, the committee realized that to avoid possible shortages it needed to find other reliable suppliers. Hence, with the tacit approval of the Kaohsiung Cold Storage 1st and 2nd Plants, vessel owners also cooperated with the Yuhan Ice Manufacturer and the Yurong Ice Manufacturer. The arrangement was similar to the one with the Kaohsiung Cold Storage 1st and 2nd Plants. While these negotiations were underway, the Hechun Fisheries Ice Manufacturer lowered its price to try to lure vessel owners away from a possible agreement and in the process to undermine solidarity among the owners.[15]

To ensure that the cooperation between the fishing companies and the ice makers worked well, the parties established a system for the fishing companies to follow in purchasing ice. They initially had to buy from Yurong, then from the Kaohsiung Cold Storage 1st and 2nd Plants and finally from Yuhan. Due to their different levels of productivity, however, the Kaohsiung Cold Storage 1st and & 2nd Plants provided the most ice, followed by Yurong and then Yuhan.[16]

The Fishing Guild Archives indicate that from 1962 onwards, vessel owners did not discuss the issue of ice manufacture and supply in their internal meetings as frequently as in the past. This was partly because they had had set up a reliable cooperative system with the ice manufacturers and partly because large refrigerators were increasingly being installed in both pair-trawlers and longliners. In 1968, Kaohsiung's fishing industries once again considered a plan to build an ice-manufacturing plant in the Chianjhen District. This idea was encouraged by the completion of Chianjhen Fishing Port, but the project was never started because modern ice-manufacturing equipment, such as huge air conditioners and refrigerators, required high import duties which forced vessel owners to drop this unduly expensive idea.[17]

---

[15]KPTOC, "4th Meeting Records of the KPTOC in 1962," 23 April 1962; "5th Meeting Records of the KPTOC in 1962," 14 May 1962; and "7th Meeting Records of the KPTOC in 1962," 16 July 1962.

[16]*Ibid.*, "5th Meeting Records of the KPTOC in 1962," 14 May 1962; "7th Meeting Records of the KPTOC in 1962," 16 July 1962; "9th Meeting Records of the KPTOC in 1962," 24 September 1962; and "10th Meeting Records of the KPTOC in 1962," 27 October 1962.

[17]KFBCG, "2nd Conference Records of the 2nd Cohort Member Representatives," 22 January 1968; and "1st Temporary Conference Records of the 3rd Cohort Member Representatives," 22 August 1970.

The Kaohsiung Fishing Industry and the Shipyards

Shortly after the dispute with the ice manufacturers abated, a new challenge arose from the shipyards over dry-dock fees. Actually, the issue was not entirely new. As far back as 1957 the Taiwan Shipbuilding Association had raised the fee [*Shangjiafei*] with the permission of the FAA. During this period of price fluctuations, the dry-dock fee in Kaohsiung Fishing Port, which previously was lower than in Keelung, was raised to the same level. Kaohsiung's fishing companies had no choice but to accept this increase.[18]

Between 1951 and 1962, as the exchange rate with the US dollar fell dramatically from NT $10.30 to NT $40.00, Kaohsiung's shipbuilding industry frequently raised the dry-dock fee on the pretext of the relentless increase in inflation. In 1962, for example, the Taiwan Shipbuilding Association took measures to increase it by sixty percent. Again, no objections to this hike can be found in the Guild Archives or the newspapers, so it appears that vessel owners acquiesced to this major adjustment. By 1968, however, the shipyards announced plans to increase dry-dock fees once again. But this time the Fisheries Bureau canvassed vessel owners and found that most believed that there was no room for a further increase. In part this was because the charges had already been raised that year, but the vessel owners also noted the continual decline in fish prices over the previous decade.[19]

Despite these objections, the dry-dock fee continued to rise over the next several years. Fishing companies seem to have tolerated it as long as the increases were not huge. At the end of 1972, however, the latest projected rise in dry-dock fees eventually strained the vessel owners' patience to the breaking point. They submitted the dispute to arbitration by the Administrative Section of Commodity Prices at the Ministry of Economic Affairs [*Jingjibu Wujia Guanli Siaozu*], stating that before making any further adjustments to dry-dock fees, shipyards should first have to consult with vessel owners. The average

---

[18]*China Fisheries News*, 7 October 1957. It was also agreed by various parties that fishers and fishing companies be allowed to rent dry-docks at shipyards and hire shipwrights to maintain their vessels. No complaints from the fishing industry about this increase were found in fisheries newspapers and magazines.

[19]Chiu, "Money and Financial Markets," 123; KPTOC, "4th Meeting Records of the KPTOC in 1962," 23 April 1962; and KFBCG, "20th Combined Meeting Records of the 2nd Cohort Directors and Supervisors," 30 September 1968. The shipbuilding industry used to give a twenty percent discount on dry-dock fees, but this preferential treatment was cancelled in 1968. Vessel owners considered this in effect a rise in the fees.

exchange rate had remained stable for more than a decade, so the vessel owners contended that shipyards should not be permitted to raise fees again.[20]

Unfortunately for the vessel owners, the government ignored their appeal, and dry-dock fees continued to increase.[21] In August 1974, the Kaohsiung shipbuilding industry arbitrarily doubled the levy. Before vessel owners could respond to this "surprise attack," Mai Cinggang, the chief executive of the Sanyang Shipbuilding Plant, took the initiative to talk to vessel owners. He proposed that as a kind of "after-sales service," vessels built by Sanyang or its allied shipbuilding yards (Haijhuan, Taiji and Fongguo) would be exempt from the new fee hike. All others, however, would have to sign a two-year contract with Sanyang if they were interested in availing of this offer.[22] Mai's suggestion, however, was considered discriminatory by vessel owners. The Guild, on behalf of the owners, then opened negotiations with the shipbuilders. It first asked for a thirty-five percent reduction in the dry-dock fee which, not surprisingly, was resolutely rejected by Mai. In September, the fishing industry asked the Kaohsiung city government and other relevant authorities to arbitrate the dispute, but none of these bodies was prepared to become involved. To extricate themselves from what they considered to be financial "extortion" by the shipyards, the vessel owners considered building their own yard, but this plan was quickly shelved due to a lack of vacant land in Kaohsiung Fishing Port.[23]

In 1975 the shipyards unilaterally doubled the charge for the dry-dock leases. After a protracted series of negotiations, the shipbuilders finally agreed to reduce this increase by a small margin. Vessel owners, however, still felt aggrieved, and the next year they struck back by suggesting a new fee system. On their behalf the Guild proposed that shipowners pay a stipulated fee, regardless of vessel size. It suggested as well that different types and sizes of

---

[20]KFBCG, "21st Meeting Records of the 4th Cohort Directors," 1 December 1972; and Chiu, "Money and Financial Markets," 123.

[21]No further information was found in the Guild achieves or fisheries magazines regarding the intervention of the Administrative Section of Commodity Prices in 1972 and 1973.

[22]KFBCG, "Symposium Records of Guild Directors and Supervisors," 26 August 1974. Sanyang, Haijhuan, Taiji and Fongguo were the most important shipbuilders at that time.

[23]KFBCG, "11th Meeting Records of the 5th Cohort Directors," 18 September 1974; "Agenda of the 11th Meeting Records of the 5th Cohort Directors," 18 September 1974; "3rd Conference Records of the 5th Cohort Member Representatives," 28 February 1975; and "Implementation Report of the Resolutions Offered by the 1st, 2nd and 3rd Directors' Meetings," 20 June 1973. Shipyards only gave vessel owners a ten percent discount on the dry-dock lease.

vessels should be able to stay in dry-dock for varying periods of time, although this could be extended if the maintenance process was interrupted by power failure, public holidays or bad weather. To encourage the shipyards to accept the new fee structure, the Guild promised to collect dry-dock fees on the shipyards' behalf to ensure that they would avoid potential disputes with the fishing companies and vessel owners. Parallel with these negotiations, the Guild also sought two or three shipbuilders with which it could cooperate to end the collusion in the shipbuilding industry – just as it had done with ice manufacturers in the 1960s.[24]

Despite its best efforts, the Guild discovered that none of the shipyards wanted either to negotiate a new fee structure or to enter into a cooperative agreement with the vessel owners. As a result, the vessel owners decided to alter their strategy to concentrate their opposition on the Sanyang Shipbuilding Plant – the initiator of the collusion scheme. They therefore decided to talk to Sanyang directly and threatened to impose a boycott if it rejected their proposals. Simultaneously, the Guild submitted the dispute to the government for arbitration.[25]

The latter strategy went nowhere when the government once again ignored the vessel owners' appeal to intervene. Hence, talking to Sanyang directly became the primary strategy, and Sanyang and its allied shipyards were soon forced to return to the negotiating table. At the end of 1976, the two parties reached a series of agreements. First, the dry-dock fee was to be fixed at twenty percent below the price that the Ministry of Transportation and Communication had set in February 1974. Vessels of 250 gross tons or less were to be allowed to stay in dry-dock for four days, while vessels of more than 250 tons could remain for up to five days. Vessels would not be allowed to leave the shipyard before vessel owners paid the dry-dock fee. If the transaction was paid for by cheque, it would have to be able to be cashed within three months. If the service period in the dry-dock extended beyond the allocated time limit, vessel owners had to pay an additional twenty-five percent of the basic fee per day for steel-hulled vessels and thirty percent per day for wooden-hulled craft.[26] Both parties had made major concessions, and the tense atmosphere was finally relieved. But there is one important point to note here:

---

[24]*Ibid.*, "Implementation Report of the Resolutions of the 1st Temporary Conference Records of the 5th Cohort Member Representatives," 28 February 1975; "3rd Temporary Meeting Records of the 6th Cohort Directors," 3 November 1976; and "6th Meeting Records of the 6th Cohort Directors," 7 December 1976.

[25]*Ibid.*, "6th Meeting Records of the 6th Cohort Directors," 7 December 1976.

[26]*Ibid.*, "4th Temporary Combined Meeting Records of the 6th Cohort Directors and Supervisors," 15 December 1976.

the government consistently remained neutral and never offered to assist the fishing industry, even though the vessel owners repeatedly requested its intervention.

Fishing Companies, Stevedores, Trucking Companies and Security Guards

Fishing owners unloaded their catch shortly after their vessels returned to Kaohsiung Fishing Port. This job required a large number of port workers known as fish stevedores [*Sieyu Gongren*]. Unlike other stevedores who offered their services to all sorts of ships, fish stevedores only unloaded fishing vessels. They had to perform their strenuous work in the early mornings before Kaohsiung Fish Market opened or before the trains and trucks left for other markets around Taiwan. Their manual job, although technically simple, was indispensable to the industry.

The fish stevedores played an important role in the smooth operation of the fishing industry. They also created many labour problems. They always asked for free fish from fishing masters and vessel owners. Given that large amounts of fish were unloaded every day, the stevedores believed it was their right to ask for one or two fish. This kind of daily request put fishing masters and vessel owners in an awkward position. In mid-1964 this sensitive issue was referred for discussion by the Guild, whose members unanimously decided not to provide free fish to the stevedores as had been the custom in the past.[27]

The fish stevedores' irresponsibility and lack of discipline also raised the ire of vessel owners. Dockside thieves mingled with the stevedores and stole fish while the catch was being unloaded. Stevedores, however, rarely bothered to stop them from pilfering. To keep dockside thieves away from the ships, vessel owners insisted that stevedores wear identification tags containing both their name and photo. They also asked the stevedores' union to replace incompetent or irresponsible foremen.[28]

Why did the stevedores not file reports when fish was stolen? The answer is simple: the stevedores also stole fish. The financial losses caused by pilferage by the fish stevedores were even higher than those caused by dockside thieves.[29] When the latter stole fish, the companies asked for help

[27]KPTOC, "7th Meeting Records of the KPTOC in 1964," 29 June 1964.

[28]KFBCG, "1st Meeting Records of the 1st Cohort Directors," 30 April 1965.

[29]Theft at the dockside, especially by stevedores, was mentioned frequently in the Guild archives. Theft by stevedores to supplement their incomes has been a common crime in ports around the world. For example, in 1920 an Australian Royal Commission estimated that in the port of Fremantle pillaging led to the loss of cargo valued at £21,760. See Malcolm Tull, "Fremantle, 1890- 1990," in Sam Davies, *et al.*

from the market security guards; however, when fish stevedores stole fish, the companies had to call the police to supervise the unloading of the catch.[30]

The stevedores used every conceivable means to steal fish, and vessel owners could not solve the problem. As a result, the relationship between the two parties was strained. The theft problem eventually reached the limit of the vessel owners' tolerance in late 1967. In that year the foremen system had not functioned well; the quality of the stevedores' work was abysmal, and the problem of theft became too much to bear. In their internal meetings the vessel owners referred to the fish stevedores as "malignant tumours," and the records in the Guild archives reveal the extent of their anger.[31]

The stevedores used a variety of techniques and strategies to steal fish. They entered the holds, stuffed fish into their galoshes or the crotch of their trousers and secretly took their booty home after work. Furthermore, because they worked in a fishing port, they knew the value of various species, so the fish they stole was almost always the most expensive.[32]

When the stevedoring crews were short of hands, the union recruited children as relief workers. On several occasions, the Guild lodged official complaints about this practice, but the union simply ignored them. To protest against the employment of child labour, the Guild demanded reforms within a specified period of time. Simultaneously, the Guild began to think about the possibilities of recruiting an alternative group of stevedores to replace the current workforce. This strategy successfully raised concerns among the stevedores about their future job security. To ease the tensions, the union promised to discipline and reform the stevedores. They did not fulfil this promise, however, and the performance of the stevedores remained poor.[33]

(eds.), *Dock Workers: International Explorations in Comparative Labour History, 1790-1970* (2 vols., Aldershot, 2000), II, 480.

[30]KPTOC, "7th Meeting Records of the KPTOC in 1964;" and KFBCG, "3rd Combined Meeting Records of the 1st Cohort Directors and Supervisors," 14 January 1966. Vessel owners asked the police to supervise the unloading of fish on the spot for one month.

[31]KFBCG, "11th Meeting Records of the 1st Cohort Directors," 28 February 1975; and "10th Combined Meeting Records of the 2nd Cohort Supervisors and Directors," 30 October 1967.

[32]*Ibid.*, "Motions Proposed in the 2nd Meeting of the 2nd Cohort Member Representatives," 22 January 1968.

[33]*Ibid.*, "10th Combined Meeting Records of the 2nd Cohort Supervisors and Directors," 30 October 1967; and "11th Meeting Records of the 2nd Cohort Directors," 30 November 1967.

Although the level of performance by the stevedores was far from satisfactory, they demanded pay raises in April 1967, August 1968 and April 1969. At the beginning of the 1970s, they were still trying to pressure the fishing companies for further wage increases. In 1974, they had the temerity to ask vessel owners for a fifty percent hike.[34] We can make a couple of generalizations about the interactions between the owners and the stevedores. First, when the two sides negotiated over pay, the stevedores liked to use the alleged improvement of their work performance as a bargaining chip. They invariably threatened to refuse to increase their productivity unless vessel owners were prepared to increase their wages. In other words, the fish stevedores did not consider productivity or efficiency as requirements of their job. Second, the union agreed that anyone who was caught stealing fish three times would forego a wage increase. In other words, stevedores were tacitly allowed to steal fish "legally" at least twice, providing they did not get caught in the act a third time.

The fishing industry was never able effectively to solve the problems of rampant theft and malingering by the stevedores. Even in the mid-1970s stealing and poor job performance remained major problems. Nonetheless, their union continued to demand pay increases as often as before.[35] Hence, it is easy to understand why vessel owners usually spoke in disparaging tones when referring to these workers.

Once the fish was unloaded, it was usually sold at the Kaohsiung Fish Market and then transported elsewhere in Taiwan. In the 1940s and 1950s, the fish products were transported by train, but by the 1960s trucks gradually became the principal method of transport. This shift encouraged Kaohsiung's fishing companies to consider the potential for cooperation with the trucking companies. The first transport boss to enter into a partnership with the vessel owners was Jhang Wunjhang. Although this cooperation began at the beginning of 1965, it did not continue for very long, and the fishing industry was constantly threatened by collusion among the trucking companies. By mid-1967, the vessel owners began to think seriously about establishing their own trucking company. But this idea never got off the ground, and the vessel owners continued to look for trucking companies with which to cooperate. In 1969 they discovered that the Hongcyuan Transportation Company was

---

[34]*Ibid.*, "4th Combined Meeting Records of the 2nd Cohort Directors and Supervisors," 22 April 1967; "19th Combined Meeting Records of the 2nd Cohort Directors and Supervisors," 30 August 1968; "4th Combined Meeting Records of the 3rd Cohort Directors and Supervisors," 30 April 1969; "2nd Meeting Records of the 4th Cohort Directors," 7 April 1971; and "8th Combined Meeting Records of the 5th Cohort Directors and Supervisors," 19 January 1974.

[35]*Ibid.*, "3rd Temporary Meeting Records of the 6th Cohort Directors," 3 November 1976.

amenable to such an arrangement because it had purchased a fleet of new trucks and needed to alleviate the financial burden. This so-called "cooperation" was only temporary; when Kaohsiung's trucking companies launched a monopoly scheme in May 1972, Hongcyuan joined them.[36]

As a result, in June 1973 the fishing companies returned to the idea of establishing their own trucking company. This decision was based not only on a desire to avoid the constant increases in transport fees but also because the truck drivers were also in the habit of stealing fish. The level of the theft became intolerable, and the vessel owners believed that if they could run their own trucking fleet and hire their own drivers, the problem would be alleviated. In August the Guild struck a committee to plan for the establishment of its own trucking company. The Deputy Head of the Fisheries Bureau, Chen Banghao, was invited to participate in this project. The resulting plan called for the fishing industries of Kaohsiung and Keelung to purchase jointly forty refrigerated trucks and 5000 *pings* of land as a base. The committee estimated that NT $53 million (US $1.38 million) would be required. To raise the funds, the fishing industries of Kaohsiung and Keelung planned to apply to both the central and provincial governments for loans and a subsidies. Any shortfall would have to be made up by the vessel owners. The amount that each company would have to pay for the new venture would depend on the number and tonnage of their vessels.[37]

Seventy-five fishing companies agreed to participate in this project. The Guild, in concert with the Keelung fishing industry, also applied for a loan via the Fisheries Bureau from the Sino-American Joint Commission on Rural Reconstruction (JCRR). News of this proposed trucking company soon spread throughout Kaohsiung's transport industry. In order to "extort" money from the vessel owners before they could launch their new venture, Kaohsiung's trucking companies quickly instituted three fee increases. This strategy, however, only spurred the vessel owners to make greater efforts to set up their

---

[36]Gu Tingfang, interview, 19 June 2002; KPTOC, "7th Meeting Records of the KPTOC in 1964," 29 June 1964; KFBCG, "11th Meeting Records of the KFBCG in 1964," 28 October 1964; "12th Meeting Records of the KFBCG in 1964," 28 November 1964; "6th Combined Meeting Records of the 2nd Cohort Directors and Supervisors," 28 June 1967; "3rd Combined Meeting Records of the 3rd Cohort Directors and Supervisors," 28 March 1969; and "15th Meeting Records of the 4th Cohort Directors," 26 May 1972. In order to ease the impact, the Guild provisionally asked for help from Jhang Wunjhang.

[37]KFBCG, "2nd Meeting Records of the 5th Cohort Directors," 23 March 1973; "Implementation Report of the Resolutions Offered by the 1st, 2nd and 3rd Directors' Meetings;" and "5th Meeting Records of the 5th Cohort Directors," 13 August 1973.

own company. This new firm, named the Liren Refrigerated Truck Ltd., commenced operations in late 1974 with Chen Banghao as its president.[38]

For the Kaohsiung fishing industry, the main advantage of having its own trucking company was to avoid "financial extortion" by the private firms. Liren, on the other hand, as a trucking company owned and operated by the vessel owners, always remained "obedient" to the fishing industry's wishes whenever there was a dispute over transport charges. An example of this was an event in December 1974 when vessel owners complained that the fees were a bit too high. Without hesitation, Liren immediately lowered them by ten percent.[39]

The establishment of Liren Refrigerated Truck Ltd. marked a milestone in the relations between the fishing industry and the ancillary sector. The fishing companies had always threatened to establish their own onshore enterprises when they had encountered a serious conflict of interest with the ancillary industries, but none of these plans had eventuated. The appearance of Liren in late 1974 changed all this.

Besides a desire to lower costs, Liren was established to cut down on the long-standing problem of pilferage. To minimize the problem, vessel owners in the past had hired "security guards" [*Yayun Renyuan*] to escort their fish to market. The history of using these security personnel dated back to the early 1960s when trucks first became the main means of transport for fish products. The guards were recommended by the fishing companies and employed as temporary workers by the Guild. At that time, no one in the Kaohsiung fishing industry expected that the guards would band come together to form a kind of trade union that was strong enough to bargain with the fishing companies. Towards the end of 1962, the Guild issued a set of "Management Regulations for Security Guards" [*Yayun Renyuan Guanli Banfa Size*]. According to these rules, before he could be hired a security guard had to first provide a letter of recommendation from a fishing company. Once an individual obtained a position, he worked for the Guild rather than the company that had recommended him. If a fishing company went into

[38]*Ibid.*, "7th Meeting Agenda of the 5th Cohort Directors," 14 December 1973; "Working Plan of 1974," n.d.; "3rd Combined Meeting Records of the 5th Cohort Directors," 21 February 1974; and "1st Temporary Conference Records of the 5th Cohort Member Representatives," 20 August 1974.

[39]*Ibid.*, "9th Temporary Meeting Records of the 5th Cohort Directors," 5 December 1974; and "13th Combined Meeting Records of the 5th Cohort Directors and Supervisors," 24 January 1975.

liquidation, however, the security guard was automatically dismissed unless he could find another guarantor in the industry to sponsor him.[40]

Still, this system could not guarantee that all the guards were trustworthy and competent. The men were largely only semi-literate and could not communicate adequately with the staff when they delivered fish products to local markets. Their level of performance was a serious concern to both the vessel owners and the market staff. Furthermore, some guards also stole fish. The first such scandal was exposed in April 1963, only five months after the "Management Regulations" were promulgated. Noting the seriousness of the problem, the Guild warned the guards that anyone caught stealing would be dismissed immediately.[41]

The performance of the security staff, similar to the experience with the stevedores, fell short of the fishing companies' expectations. These workers frequently asked the Guild for wage increases. In 1973, for example, the guards demanded pay hikes in both March and December, claiming that the cost of living in Kaohsiung had increased substantially. For whatever reason, the Guild assented.[42] But in April 1974 the security guards were planning to establish a union. This radical step surprised the vessel owners, who wrote letters of complaint to the relevant authorities, including the Kaohsiung city government, the Kaohsiung City Police Bureau, the Ministry of the Interior and the military.[43]

Their lobbying, however, did not work. The establishment of the union was already underway before the city government could seriously examine the draft of the union's regulations. To canvass opinions from the fishing industry, the city government sent a copy of the draft to the Guild in July 1974. The Guild's refused to read it and instead sacked all the guards.[44]

---

[40]KPTOC, "Records of Temporary Meeting in November of 1962," 10 November 1962; "2nd Meeting Record of the KPTOC in 1963," 10 March 1963; and "1st Temporary Meeting Record in December 1962," 18 December 1962.

[41]KFBCG, "2nd Conference Records of the 1st Cohort Member Representatives," 27 January 1966; and KPTOC, "3rd Meeting Record of 1963," 3 April 1963.

[42]KFBCG, "2nd Meeting Records of the 5th Cohort Directors," 23 March 1973; and "7th Meeting Records of the 5th Cohort Directors," 14 December 1973. The Guild approved their request in principle and authorized the director of the marshalling station [*Diaopeijhan*] to negotiate with the representatives of escorts concerning a wage increase.

[43]*Ibid.*, "5th Temporary Meeting Records of the 5th Cohort Directors," 17 April 1974.

[44]*Ibid.*, "6th Temporary Meeting Records of the 5th Cohort Directors," 23 July 1974; and "1st Temporary Conference Records of the 5th Cohort Member Repre-

Fishing Equipment Suppliers

Fishing vessels required ropes, steel cables, fish boxes, tackle and nets for every offshore and long-distance voyage.[45] Such equipment traditionally was acquired from a variety of factories or shops. In the immediate post-World War II era, the price of fishing equipment was not a major concern, and the relationship between the Kaohsiung fishing industry and their equipment suppliers was harmonious.[46] This amicable situation continued until the suppliers increased the price of steel cable in April 1972. This sudden price rise placed a considerable financial burden on the fishing companies, which charged the suppliers with collusion. On behalf of the vessel owners, the Guild lodged a formal complaint with the Bureau of Foreign Trade [*Guomaojyu*]. In addition two solutions were proposed at Guild meetings. First, it was suggested that steel cables be imported directly from abroad, an approach that the Keelung fishing industry had used for a while. Second, it was proposed ed that the companies establish a plant to produce their own cable. The former required the government's approval, while the latter seemed to many to be creating a huge problem over what was really a small matter. Ultimately, neither captured the vessel owners' interest.[47]

A year later the Kaohsiung fishing industry once again asked the Fisheries Bureau and the Ministry of Economic Affairs to intervene in a dispute with the equipment suppliers. The authorities, however, remained neutral because neither the vessel owners nor the police could find any evidence of collusion.[48] The continuing price rises, however, raised concerns throughout Taiwan. Hence, in July 1974 the Taiwan Provincial Fishermen's Association suggested establishing a publicly-owned "Fishing Equipment Supply Cooperative" [*Yuyong Wuzih Hezuoshe*]. It planned to draw in part on

---

sentatives," 20 August 1974. Once the proposed Liren Truck Company started operations, they would no longer need security guards.

[45]Fishing ropes are more durable than ordinary ropes. They are produced for fishing purposes only and can be used in the water for long periods of time. Fish boxes are designed to hold the catch; they are used largely on trawlers.

[46]None of disputes between the Kaohsiung fishing industry and fishing equipment suppliers can be found for the early period in the Guild Archives or newspapers.

[47]KFBCG, "14th Meeting Records of the 4th Cohort Directors," 1 April 1972; "15th Combined Meeting Records of the 4th Cohort Supervisors and Directors," 26 May 1972; and "17th Combined Meeting Records of the 4th Cohort Supervisors and Directors," 26 July 1972.

[48]*Ibid.*, "Implementation Report of the Resolutions Offered by the 1st, 2nd and 3rd Directors' Meetings."

a NT $10 million subsidy from the central government, with the rest of the expenses being shared among the fishing guilds in Kaohsiung and Keelung and other fishermen's associations in Taiwan. This project certainly interested the Kaohsiung vessel owners, who even proposed a fundraising scheme. In the end, though, the cooperative never materialized.[49]

To cope with the emerging equipment supply problem, at the end of 1974 the Guild established three small committees to seek cooperative arrangements with dry-dock, rope and net, cable and fish box suppliers who had not participated in any form of collusion.[50] But no matter how much effort the committees expended, they could not fully eliminate the problems created by the suppliers; the price of fishing gear continued to increase while any cooperation between the fishing industry and its suppliers could end at any time. The vessel owners patiently repeated their previous tactics, and at the end of 1976 the Guild for a third time submitted the dispute with the suppliers to arbitration by the city government and the Kaohsiung City Police Bureau.[51]

Compared with the challenges posed by other ancillary industries, the threat from the equipment suppliers was relatively small, in part because the majority were really cottage industries. Moreover, their relative importance to the fishing industry was less than the other large-scale, modern enterprises, such as the shipyards or the trucking companies. Thus, the vessel owners were not as concerned as when they faced the threat of collusion from the ice makers, shipyards or transport companies.

---

[49]*Ibid.*, "10th Meeting Records of the 5th Cohort Directors," 6 July 1974; "1st Temporary Meeting Records of the 5th Cohort Representatives," 20 August 1974; and "11th Meeting Records of the 5th Cohort Directors," 18 September 1974. In September, the Guild leaders decided to postpone the fundraising indefinitely. See KFBCG, "Symposium Records of Guild Directors and Supervisors," 26 August 1974. Many things were happening at the time. In August, the Kaohsiung shipbuilding industry arbitrarily doubled the dry-dock fees. The threat from shipbuilders was more serious than that from the fishing equipment suppliers; KFBCG, "1st Temporary Conference Records of the 5th Cohort Member Representatives," 20 August 1974. Also, vessel owners were heavily preoccupied with the establishment of Liren Truck Company during this period.

[50]KFBCG, "9th Temporary Meeting Records of the 5th Cohort Directors," 5 December 1974; and "13th Combined Meeting Records of the 5th Cohort Supervisors and Directors," 24 January 1975. In a very brief period of time they had successfully sought out the Gangshan Factory of the Veterans Affairs Commission (VAC) with whom to cooperate. The VAC was a government unit which ran a variety of businesses to support veterans. Hence, the ropes they produced were cheaper than those of private rope makers.

[51]*Ibid.*, "5th Meeting Records of the 6th Cohort Directors," 19 October 1976.

When the vessel owners had business disputes with the shipyards and ice makers, they always discussed how best to retaliate and fought together until the threat was minimized or dissipated. But this was not the case when they dealt with the demands by equipment suppliers. The vessel owners talked about equipment supply problems once or twice, but then did not mention it again for several years. Clearly, they had far more pressing matters on their minds. The Guild Archives provide an example of this ambivalent attitude. In April 1973, when they asked the government to mediate their dispute with the equipment suppliers, the request was dutifully ignored, and it did not seem to worry the vessel owners. They did not discuss the equipment supply issue again until the Taiwan Provincial Fishermen's Association suggested setting up an equipment cooperative in July 1974.[52] In the interim, vessel owners were heavily preoccupied with more important issues, such as their relationships with the trucking companies and the shipyards. Dealing with the threats from the equipment suppliers was not a top priority. This may in part explain why the plan to set up a Fishing Equipment Supply Cooperative was dropped.

Moreover, the number of shipyards, trucking companies and ice manufacturers in Kaohsiung was limited, so when they tried to collude, the fishing industry could readily identify the key players and mount a counterattack against that person and his allies. But since there were a plethora of fishing gear shops in the port, finding the leader was almost impossible. Given this reality, it is easy to understand why the owners had no real idea against whom to retaliate. This can also explain two important aspects of the problem: first, why the fishing industry always relied on the traditional approaches it had used in the past; and second, why the government never took any effective steps to mediate the chronic disputes between the fishing companies and their suppliers. Neither the vessel owners not the government knew who actually organized the collusion and hence could not figure out with whom to negotiate.

A closer look at the way that the Kaohsiung fishing industry dealt with other industries during disputes also reveals several other things. In the first instance, vessel owners to some degree were concerned only with their own self-interest. This attitude was clearly demonstrated in two tactics. First, they always asked the ancillary industries to lower their prices in order to reduce their own costs and hence maximize their profits. The most common reason they gave was that the price of fish had continued to decline over the years and thus any increases in service charges would hinder the future development and well-being of the industry. This kind of economic argument was not persuasive enough to convince the other industries to agree because

---

[52]I have read every page of the Guild records carefully from April 1973 to July 1974. None of the vessel owners mentioned the issue of fishing gear supply, even though their business disputes with gear suppliers remained unsettled.

the development of the fishing industry was not their sole concern. Hence, they did not feel obliged to sacrifice their own business interests for the sake of the fishing companies. Second, while the fishing vessel owners always took collective action in their struggles against other industries, they never accorded their foes the same right. Instead, any "joint action" by the ancillary industries was always construed as a form of collusion.

The collective actions taken by the vessel owners usually exhibited two features. Internally, they tended to establish provisional committees within the Guild to lobby government, negotiate with their rivals or mobilize the other vessel owners. These committees could be reorganized at any time in accordance with the specific requirements of a situation and then disbanded when they had fulfilled their tasks. Moreover, through a coordinated division of labour, the fishing industry was able to deal with a range of threats simultaneously.

Externally, vessel owners in Kaohsiung learned how to form alliances with their counterparts in Keelung whenever they confronted a common threat. This collaboration was strengthened after the first supplementary legislator and National Assembly delegate were elected by the fishing communities in 1972. Luo Chuanjin, a National Assembly delegate from Kaohsiung, also had to speak for the Keelung fishing community, while Huang Zecing, a legislator from Keelung, had to attend Guild meetings in Kaohsiung.[53] Thus, the collective actions of the vessel owners were in effect nationwide. The scale and impact of their activities was consequently much stronger than any attempt at collusion by the ice manufacturers or shipbuilders.

There are several other points that need to be emphasized. When the fishing industry took collective action, not all vessel owners were necessarily keen to participate. Some enterprises, as we saw in chapter four, engaged in a variety of ancillary businesses in Kaohsiung, including ice manufacturing, shipbuilding and trucking. Their separate but related business interests were certainly different from the other vessel owners, and this was especially noticeable whenever there was a conflict. The Hechun Fishing Company provided a good example of these conflicts of interest. Although Hechun was one of the most successful fishing companies in Kaohsiung, when the vessel owners took on the ice manufacturers, its ice-making department, the Hechun Fisheries Ice Manufacturer, did not support them. Instead, Hechun tried to undermine effective cooperation between the Guild and other ice makers and made life difficult for other members of the Kaohsiung fishing industry.[54]

---

[53]See chapter 4.

[54]The rise in ice prices would never affect the running of the Hechun Fishing Company because it could always get a sufficient supply from its ice-manufacturing department. The same situation also occurred in the disputes between the fishing and shipbuilding industries.

Furthermore, as we have seen, government did not always side with the fishing companies in their disputes with the ancillary industries. Disagreements between the shipyards and the fishing companies in the 1970s provide good examples. When vessel owners asked for help in 1972, 1974 and 1976, the government did not intervene as it had in the past. As a consequence, the fishing industry was forced to beat a retreat. Basically, the struggle between the vessel owners and the shipbuilders was an economic battle between a primary and a heavy industry, and the government recognized that the importance of the former could not compare with the significance of the latter.

When other industries pursued collusive practices against the fishing companies, there were always some men who chose to "betray" their colleagues and enter into partnership agreements with the vessel owners. The Jhumao Shipbuilding Plant provides another example. When the shipyards raised dry-dock fees in 1964, Jhumao chose to offer the fishing companies a twenty percent discount in order to increase its market share.[55] This kind of competition offered vessel owners the opportunity to thwart the attempts by the other industries to create monopolies. The vessel owners also learned the hard way that such cooperation seldom lasted for an extended period of time.

Finally, their interactions with workers from other industries clearly demonstrated that the vessel owners treated people of various social classes differently, both in terms of their attitude and behaviour. When the vessel owners dealt with the ice manufacturers, shipyard owners or trucking company bosses, they had to behave amicably. But this was not the case when they dealt with working-class people like security guards and fish stevedores. On the one hand, the vessel owners could be paternalistic and often agreed to raise the wages of the workers; on the other hand, however, they could also be arrogant and even harsh, especially when the workers' performance did not meet their expectations. A classic example was the decision in July 1974 to dismiss all the security guards rather than seek a way to ameliorate the dispute.

---

[55]KPTOC, "3rd Meeting Records of the KPTOC in 1964," 30 March 1964.

Chapter 6

The Development of Kaohsiung's Fishing Companies

After the Second World War there were dramatic changes in the character of the Kaohsiung Fishing Port. Besides the construction of major facilities and the development of ancillary industries, the ethnic composition of Kaohsiung's fishing industry differed markedly from the colonial era on two important grounds. First, the Japanese had been expelled, and their companies and fisheries-related enterprises were taken over by the government of the Republic of China (ROC) in 1945. Second, several waves of migration from various parts of Taiwan had provided Kaohsiung's fishing industry with a diversity of ethnic groups.[1] The Penghuans, mainland Chinese, Siao Liouciounese and native Taiwanese all played crucial roles in specific segments of Kaohsiung's fishing industry.

The Penghuans comprised the main source of labour in the distant-water fisheries, especially in the longline sector. They generally served as employees on the vessels and few ever started their own companies. The mainland Chinese, primarily from Shandong and Fujian, worked mostly for the private trawling companies and seldom entered the longline fishery. Because of this, they were greatly affected by the implementation of the 200-mile Exclusive Economic Zone (EEZ). Most disappeared from Taiwan's fishing industry after international fishing cooperation ended in the 1980s. Most of the fishers from Siao Liouciou stayed in the longline fishery. They invariably started their careers in offshore longlining in the waters off Siao Liouciou and gradually extended their fishing into Southeast Asian waters. To survive in Kaohsiung's metropolitan environment, the Siao Liouciou people formed a self-contained and closed community; each member, from vessel owners to fishing masters and apprentices, was a kinsman. The conservative nature of the fishers from Siao Liouciou has to a degree limited their potential in the global fisheries.

The local Taiwanese from Kaohsiung, on the other hand, never concentrated on any particular fishing sector. Most restarted their businesses by operating longline fishing companies after World War II, but they also made attempts to launch other activities. Hence, their capital was sometimes

[1] Sia Fujhong, interview, 24 April 2004.

invested in trawlers, sometimes in longliners, sometimes in the offshore fisheries and sometimes in the distant-water sector. If they were able, they often extended their business interests into the ancillary industries, such as shipbuilding and ice manufacturing. A wide range of investments not only improved their chances of survival in an erratic business environment but also enhanced their capability to engage successfully in the global fishing industry of the postwar era.

Each ethnic group, except the Penghuans, established their own fishing companies in Kaohsiung. While all started from the same point, their responses to global challenges varied. To understand this, in this chapter I will narrow the focus to the company level. Through the histories of various companies, I will examine the business environment of Kaohsiung's fishing industry and the trajectory of fisheries development, with particular reference to changes in the grounds exploited and the improvement of fishing technology. Four representative fishing companies will be highlighted.

The Development of Trawl-Fishing: Gu Tingfang and Lin Changren

The large-scale migration of trawlers from China in the late-1940s contributed directly to the revival of Taiwan's trawl-fishing industry.[2] Initially, the Chinese vessels were centred in Keelung, but they moved south to Kaohsiung to exploit the fishing grounds off Southeast Asia (see figure 6.1). To illustrate how trawler owners developed their businesses, I will examine two vessel owners in depth. One is Gu Tingfang, who came originally from Shandong, and the other is Lin Changren from Fujian. Both were well-known in Kaohsiung's fishing industry and are ideal representatives of their communities.

Gu Tingfang was born in Shandong, a province in northern China that abuts the Yellow Sea. His family had made a living from trawl fishing for several generations, but the triumph of the communists drove the Gu family to Shanghai and then to Hong Kong. Finally, they decided to relocate to Keelung. Unlike most postwar refugees, the Gu family was wealthy, and they brought several trawlers – *Chengsinli 1 and 2*, *Huafu 1 and 2*, *Huayi 1 and 2*, *Huafong* and *Cyunli* – and a large number of crew to Keelung. These vessels and their loyal crews not only helped to maintain the family's social and economic status but also gave them the confidence to restart their successful fishing business in Keelung.[3]

In the early postwar years, it was relatively simple to operate a trawl-fishing company. The vessel owner normally was also the managing director.

---

[2]Wu Mingzuo, "A Brief Introduction to the Development of Taiwan's Trawl Fishing Industry" (unpublished Kaohsiung Fishing Boat Commercial Guild [KFBCG] internal document, n.d.), 1.

[3]Gu Tingfang, interview, 7 April 2002.

Ordinarily, a company had several staff members – an accountant, marketing manager and vessel manager. The young Gu Tingfang took immediate charge of both marketing and vessel maintenance when his family settled in Keelung. When the catch was unloaded each afternoon, Gu would arrange to have the fish transported by train across Taiwan the same night. The Jhongyang Fish Market in Taipei was the most important destination, while the remaining fish were sent to other markets along the rail line. If everything went according to plan, the fish, still in a fresh state, would arrive at Kaohsiung early the next morning. The Gu family quickly found that the transport system and marketing channels for fish were more efficient in Taiwan than in China.[4]

Figure 6.1: Mainland Chinese Fishing Migration: China-Keelung-Kaohsiung

Source: Courtesy of the author.

But the situation in Taiwan did not prove to be nearly as simple as the Gu family had originally thought. The problems emerged gradually. First, their fleet had operated out of China for at least twenty years, and the cost of maintenance was high. Second, the new fishing grounds off Taiwan were not

[4]Gu, interview, 19 June 2002.

totally familiar to their crews. As a result, the average catch was disappointing. To economize on operating costs, they decided to retire their oldest vessels, *Chengsinli 1 and 2*. When this strategy did not improve their situation, they retired their remaining vessels over the next two years.[5]

In 1951, however, the US began to provide Taiwan with aid.[6] With a long history in the fishing business, the Gu family's application for American assistance was accepted without reservation by the Fisheries Administration Agency (FAA). As a result, in 1951 the Gu family built a unit of modern pair-trawlers, *Shengchang Nos. 1 and 2*. These two vessels was not only the first time in postwar Keelung that a private company owned steel trawlers but also the initial experiment with diesel engines. The construction of the two *Shengchang*-class trawlers caused a sensation in Keelung.[7]

To operate such modern vessels required proficient fishing masters. Unfortunately, none of the Gu family's old crew knew how to fish on the new grounds or how to handle such modern vessels. The only option therefore was to hire a fishing master from overseas. The implementation of the "Regulations Concerning the Importation of Motorized Fishing Vessels of Taiwan" [*Taiwansheng Guowai Dongli Yuchuan Shuru Guanli Banfa*] in early 1951 enabled them to hire a Japanese fishing master with considerable experience on the fishing grounds off Taiwan. Yet not even the use of modern vessels and the employment of a seasoned Japanese fishing master brought success. Although he had excellent credentials, the Japanese fishing master died unexpectedly after eating a poisonous blowfish; this forced the family to rehire their old crew. The results, however, proved far from satisfactory: running costs continued to increase, and the financial burden soon became unbearable. As a result, in 1954 the Gu family was forced to sell *Shengchang 1 and 2*. The subsequent closure of their business was a difficult decision, especially since the returns from the sale of the two ships could not begin to repay the debts they had incurred in trying to re-establish their family business in postwar Taiwan.[8]

His previous experience in marketing and vessel management benefited Gu Tingfang greatly after 1951, despite the collapse of the family business. For four years he worked for various fishing companies in Keelung as a professional manager. During this period, an ever greater number of Keelung's fishing vessels were engaged in trawl fishing, but subsequent

---

[5]Gu, interview, 7 April 2002.

[6]*The Bluebook on US Aid and the Achievement of Taiwan's Fishing Industry* (Taipei, 1966), 9-10.

[7]Gu, interview, 7 April 2002.

[8]*Ibid.*

problems caused by over-fishing gradually emerged on the northern grounds. As a result, the pair-trawlers from Keelung began to exploit the southern part of the Taiwan Strait and the South China Sea; this also encouraged them to relocate to Kaohsiung in order to be closer to the fishing grounds. Gu moved there in 1958 in search of greater opportunities. Through his network of contacts with others from Shandong, Gu found a position in the Guofong Fishing Company.[9]

The head of this company, Tao Zihhou, was a business tycoon from Shandong who had invested in a variety of businesses in postwar Taipei, including a flour mill and other manufacturing endeavours, and he also attempted to operate a fishing company in Kaohsiung. Tao simultaneously built three units of pair-trawlers - *Guofong Nos. 101 and 102, Guofong Nos. 201 and 202* and *Guofong Nos, 301 and 302* - at a cost of at least NT $3 million. This huge investment caused quite a stir in Kaohsiung's fishing circles. Yet as a private company, Guofong attempted to keep the number of its onshore staff to a minimum. Only several key employees were hired. Besides the managing director, there was an accountant, vessel, maintenance and marketing managers and several couriers. All were from Shandong when the twenty-eight-year-old Gu became marketing manager. While working for Guofong, Gu witnessed a major change in the transport of fish products within Taiwan as trains were rapidly replaced by trucks. Moreover, fish were now being boxed in ice as soon as they were unloaded. With an expanding highway system, Guofong's products could easily be transported throughout the island.[10]

The first three units of pair-trawlers achieved excellent results in the South China Sea. The considerable profits and stable business environment in Kaohsiung encouraged Guofong to build other vessels. Since the fisheries authorities encouraged the development of single-trawl fishing in the late 1950s, Guofong, with financial assistance from the US and the support of the Taiwan government, built *Jhongsing No. 8* in Keelung in 1959 to operate on the northern fishing grounds. Thereafter, Guofong became one of the few companies to operate on both the northern and southern fishing grounds. After the launch of *Jhongsing No. 8*, three more units of pair-trawlers - *Minfu Nos. 1 and 2, Minfu Nos. 11 and 12* and *Tongfong Nos. 11 and 12* - were built in the next two years. All were equipped with large, modern refrigerators that could help keep the catch fresh during a long voyage. *Minfu Nos. 1 and 2* were able to fish as far away as Hong Kong, while the larger *Minfu Nos. 11 and 12* and *Tongfong Nos. 11 and 12* could fish as far south as Saigon. There are two

---

[9] *A Brief Introduction to Agricultural Industries, Vol. 10: Fisheries* (Taipei, 1962), 22; and Gu, interview, 7 April 2002.

[10] Gu, interviews, 7 April and 19 June 2002; and *China Fisheries News*, 17 April 1958.

points to emphasize here. First, in the early 1960s the majority of Taiwanese pair-trawlers still operated in the East China Sea and the northern sector of the South China Sea. *Minfu* and *Tongfong* were thus in the vanguard of the invasion by Taiwanese trawlers into the waters off Southeast Asia. Second, the average tonnage of the Taiwanese pair-trawlers gradually increased during this period from eighty tons to 150 tons, as they continued to exploit the fishing grounds in the South China Sea and as steel-hulled vessels gradually replaced traditional wooden-hulled craft.[11]

In 1964 Tao made the surprising decision to close his prosperous fishing business. He appears to have reached this conclusion partly because he wanted to concentrate on his onshore enterprises and partly because he was tired of having to travel back and forth between Taipei and Kaohsiung. *Tongfong Nos. 11 and 12* were purchased by his managing director and shareholder, Sun Shaoru, and renamed *Tongfu Nos. 11 and 12*. Sun also bought another unit of pair-trawlers, *Huafong Nos. 11 and 12*, from a different company and named his new fishing business the Tongfong Fishing Company. The reorganization of the company did not adversely affect Gu's career; he was hired by Sun as managing director.[12]

In 1966, Sun expanded his business to include longline fishing. He built *Tong'an No. 11* and *Tong'an No. 21* in 1966 and *Tong'an No. 31* and *Tong'an No. 51* in 1967. Sun named his longlining branch the Tong'an Fishing Company. In fact, Tongfong and Tong'an were divisions of the same business and were separate in name only; both were located in the same office and managed by Gu Tingfong. For Gu, the new managerial responsibility of the longlining business proved extremely challenging. Originally, the Shandong expatriates in Taiwan were not interested in longline fishing and lacked a strong business network in the industry; therefore, Gu could only obtain limited financial backing and skilled manpower. Furthermore, in the 1960s Taiwan's longline industry had expanded into a global enterprise. The operation of overseas supply bases and cooperation with Japanese and American fish dealers was a big hurdle to overcome for a monolingual group like the Shandong people.[13]

Gu managed to remove these obstacles by cooperating with the Mingtai Company and the fisheries agency of the Taiyo Fishery Co., Ltd. - an internationally known Japanese enterprise - on the condition that the entire catch be sold to Taiyo at negotiated prices. Cooperation with Taiyo greatly

[11]Gu, interviews, 7 April and 23 June 2002; and *China Fisheries Monthly*, No. 375 (1984), 25-34. The fishing grounds were extended to the southern part of the South China Sea after the mid-1960s.

[12]Gu, interview, 19 June 2002.

[13]*Ibid.*

benefited Tong'an and gave it access to the financing and port services that Taiyo provided at its overseas bases. Equally important, it meant that Tong'an would not have to develop its own international marketing channels. The overseas supply bases which Tong'an's longliner fleet now used were located at Samoa, Abidjan, Las Palmas and Cape Town.[14]

In the late 1960s, the problem of over-fishing gradually emerged in the South China Sea. The pair-trawlers could no longer catch as many fish as before, and the situation was exacerbated by the aging of Tongfong's pair-trawler fleet. In 1969, Sun decided to wind up his trawl-fishing operations to concentrate solely on longline fishing and his onshore enterprises.[15]

Gu was not interested in managing the longlining business, so two years after the closure of the trawling operations he ended his partnership with Sun and made plans to retire from the fishing industry. But the successful voyage of a Taiwanese experimental fishing vessel, *Haicing Hao*, in the waters off northern Australia in 1971 convinced Gu to return to trawl fishing. This time, though, he did not intend to be an employee but rather a vessel owner. His first pair-trawlers, *Yuanyu 11 and 12*, were built in 1973.[16] The catch rate on the Australian fishing grounds proved to be excellent (see figure 6.2). Indeed, the annual catch by Taiwanese pair-trawlers off northern Australia was 80,000 tons in 1973 and 1974. The fish holds were always filled quickly, and the maiden voyage of *Yuanyu* consumed only forty-eight days. The success of *Yuanyu* encouraged Gu in 1974 to build *Baoyuan 1 and 2* to work in the waters between Indonesia and northern Australia. In the late 1970s, however, Taiwan's trawler fleet suddenly encountered severe challenges posed by the implementation of the 200-mile EEZ. The distant-water fishing grounds used by Taiwan's pair-trawlers were legally converted overnight into the property of various coastal states. As a result, pair-trawlers were no longer able to operate as before in these areas unless they paid hefty fees. As a result, many Shandong vessel owners left the industry, but Gu decided to stay. He

---

[14]*Financial Survey of Taiwan's Fishery Enterprises of 1989* (Kaohsiung, 1991), 35; and Gu, interview, 23 June 2002.

[15]"Rambling Talk on Taiwan's Distant Water Pair-Trawl Fishing" (unpublished KFBCG internal document, n.d.), 1; Wu Mingzuo, "A Brief Introduction to the Development of Taiwan's Trawl Fishing Industry" (unpublished KFBCG internal document, n.d.), 1; and Gu, interview, 23 June 2002.

[16]"Running of Trawl Fishing" (unpublished KFBCG internal document, n.d.), 2; and Gu, interview, 3 June 2002. Many local and Shandong vessel owners built pair-trawlers in 1973. See *Financial Survey*, 34.

recognized the importance of international fisheries cooperation agreements, paying fees to purchase fishing rights in bountiful foreign waters.[17]

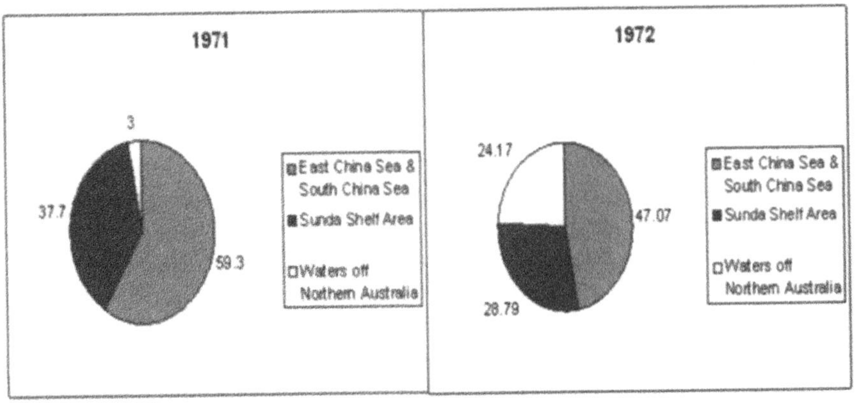

Figure 6.2: Proportional Change in the Catch of Taiwan's Distant-Water, Pair-Trawl Industry in the East and South China Seas, the Sunda Shelf and Northern Australia, 1971-1972

Source: National Taiwan University, Institute of Oceanography, "Annual Report of Catch Statistics by Area on Taiwan Demersal Fish Fisheries," 1971-1972.

Lin Changren, the second case under consideration, was born in 1924 in Fujian, a coastal province in southeast China. He was educated at the Jimei Fisheries Vocational School in the early 1940s. In 1949, Lin was sent to work for Shanghai's Jhonghua Fishing Company as a trawling apprentice. But he only remained in Shanghai briefly until severe social unrest caused by the civil war drove him to Keelung. Separated from home and family connections, Lin's only desire was to survive. Fortunately, circumstances on the island were not as miserable as he had expected.[18]

When he arrived, the Fisheries Branch of the Agriculture and Forestry Company of Taiwan (AFCT) was attempting to resurrect Keelung's trawl-fishing industry. A large number of trawl fishers were urgently needed, and with his training Lin quickly found a job at the Fisheries Branch, where he was promoted to fishing master two years later. There were plenty of job opportunities created by Shandong migrants awaiting him in Keelung. Over the next several years, Lin moved from one trawl-fishing company to another. He

[17] *China Fisheries Monthly*, No. 375 (1984), 28; and Gu, interview 23 June 2002.

[18] Lin Changren, interview, 17 May 2002.

was surprised to learn in the course of moving between companies that only a few Japanese fishing masters remained in Taiwan.[19]

In those days, most trawlers from Keelung operated on the East China Sea and the northern part of the Taiwan Strait. Fishers believed that the optimal operational environment for trawlers was along the continental shelf at depths of no more than sixty metres. But their trawlers were small, lacked power and could work only between Shanghai in the north and Penghu, a tiny archipelago in the middle of the Taiwan Strait, in the south. At the same time, an increasing number of vessels owned by private fishing companies were engaged in trawl fishing, and the problem of over-fishing in the East China Sea soon emerged at Keelung. Marine resources were being rapidly depleted and the catch correspondingly reduced. To ameliorate this, in 1953 Lin piloted his vessel to the southernmost waters of Penghu. Since the waters in the southern Taiwan Strait, especially directly off the Taiwan coast, are deeper, vessels of less than 100 tons were not suitable there. On the other hand, the waters off Fujian were comparatively shallow. As a result, Taiwan's trawl-fishing grounds increasingly expanded to the south. The richness of the marine resources in the southern fishing grounds attracted numerous vessel owners to move to Kaohsiung, and Lin joined them in 1955.[20]

Five years later, encouraged by the rapid growth of the distant-water fishery, Lin decided to start his own fishing company. For such a young man, buying a unit of pair-trawlers was prohibitively expensive. To overcome this problem, he invited fishing masters from among the Jimei alumni to join him in a joint venture, and he raised about NT $1.5 million in capital. As the largest shareholder, Lin invested about NT $700,000, his life savings from several years of risky sea ventures in southern waters, and he became head of the new company.[21]

Compared to other vessel owners, Lin's management style was conservative. According to the standard practice in Taiwan's fishing industry, a group of fishing masters could set up a fishing company with as little as twenty to thirty percent of the total capital, and then either obtain loans from the government or seek credit from the ancillary industries. They were aware that their level of financing was weak, but they were usually unconcerned. They believed in risk taking, and for many, running a new fishing company was akin to taking a gamble. If their companies made a profit, they secured new loans from banks and purchased more vessels. If their companies went bankrupt, they simply went back to sea as fishing masters. Lin, however, did not conduct his business in such a risky way. He and his partners had invested

---

[19]*Ibid.*

[20]*Ibid.*; and *Fisheries Tribune*, No. 58 (1966), 15.

[21]Lin, interview, 17 May 2002.

71.4 percent of the required capital, and their company was relatively strong compared to its rivals.[22]

The first pair-trawlers Lin bought were *Changchun Nos. 1 and 2*, two-year-old vessels that cost NT $2.1 million. After deducting the operating costs, the maiden voyage earned NT $2.5 million. Still, the size of the catch did not meet Lin's expectations due to the lack of adequate engine power. His 120-ton vessels had 120-horsepower engines, but for operations in the East China Sea these were barely powerful enough to trawl properly. In the deeper waters in the southern part of the Taiwan Strait or in the South China Sea, such pair-trawlers were totally inadequate, so malfunctions occurred regularly. To improve the overall performance of his company, Lin intended to purchase another unit of pair-trawlers as soon as his financial circumstances allowed. This opportunity arrived in 1961 when Lin's old boss, Lin Yujia, a local tycoon who had previously run the Taifong Fishing Company and afterwards established the Taiwan Glass Corp., sold him *Changtai Nos. 1 and 2* at a very low price.[23]

Figure 6.3: Wooden Hull and Frame of a Taiwanese Fishing Vessel in the Early Postwar Years

Source: Courtesy of Lin Changren.

Changtai Nos. 1 and 2, two 150-ton pair-trawlers, had 240-horsepower engines. With this new unit of pair-trawlers, Lin's crew was able to work in the waters off Hong Kong and in the South China Sea without difficulties. Following the rapid penetration of Taiwanese trawlers into Southeast Asian waters, the fishing grounds soon extended over the entire South China Sea. The horsepower and fishing gear of the new pair-trawlers were more than adequate for the task. In 1962, Lin bought the six-year-old

[22]*Ibid.*; and Wu Tianruei, interview, 8 April 2002.

[23]Lin, interview, 17 May 2002.

Changchun Nos. 11 and 12, 200-ton, steel-hulled trawlers powered by 400-horsepower engines. Although wooden-hulled trawlers were considered old after five years (see figure 6.3), steel trawlers could be maintained in good condition for at least fifteen. Compared to Lin's previous trawlers, *Changchun 11 and 12* were already relatively old, but they were still powerful vessels in good working order. This enabled his crews to exploit the distant marine resources in the waters off the southern tip of Vietnam.[24]

Figure 6.4: Distant-Water Pair-Trawlers *Changtai Nos. 11 and 12* in 1967

Source: See figure 6.3.

After struggling financially for several years, Lin's economic strength became stronger than ever. In 1965, he built *Changrong Nos. 1 and 2*, and *Changtai Nos. 11 and 12* followed two years later (see figure 6.4). All these trawlers were 200-ton vessels with 600-horsepower engines. In 1971, the good results of Taiwan's experimental vessel, *Haicing Hao*, off northern Australia encouraged Lin to build several more large vessels. In 1973, he launched *Changsheng Nos. 1 and 2*, 300-ton vessels with 1100 horsepower (see figure 6.5). In 1977, he bought *Changyun Nos. 1 and 2*, also 300-ton vessels with 1100 horsepower. Unfortunately for Lin, these were confiscated by the Indian Navy in 1978 for illegal fishing.[25]

---

[24]*Ibid.*

[25]"Running of Trawl Fishing," 2; and Lin Changren, interview, 23 May 2002.

Figure 6.5: Distant-Water Pair-Trawlers *Changsheng Nos. 1 and 2* in 1973

Source: See figure 6.3.

The crisis created by the introducion of the 200-mile EEZ in the late 1970s taught Lin that he could not put all his financial eggs in one basket. In addition to participating in international cooperation arrangements, he decided to diversify into distant-water gill-net fishing. After 1981, he purchased four second-hand longliners – *Changli No. 1*, *Changjhan No. 1*, *Changmao No. 1* and *Changhong No. 1* – and converted them into gill-net vessels.[26]

Operating Longline Fishing Companies: Pan San'guang and Chen Shengli

Most of Kaohsiung's vessel owners restarted their enterprises in the postwar years by establishing longline fishing companies. They fished in the waters off Southeast Asia in the 1950s and extended their operations to the Pacific and Indian oceans in the 1960s. But the exploitation of more remote grounds did not lead to a lack of activity in Southeast Asian waters, for the gap was quickly filled by offshore longliners owned mainly by local fishing families and mi-

[26]Lin, interview, 23 May 2002.

grants from Siao Liouciou. Two case studies will illustrate this process: Pan San'guang and Chen Shengli.

Pan San'guang was born in 1942 in Cihou, a business area that faced Gushan on the opposite side of the main waterway leading into Kaohsiung Port. His family had been prominent, wealthy and influential since the colonial era. They engaged in junk shipping and had branches in the port cities of Japanese-controlled China, conveying goods through the Taiwan Strait. Moreover, the Pans were one of the most important earthenware producers; using their sailing fleet, they dominated the earthenware trade of southern Taiwan. The Pans were also involved in Taiwan's coastal fisheries. As early as the 1920s they used large, non-motorized bamboo rafts to fish in the coastal waters off Kaohsiung. After the war they moved into motorized fishing vessels, and by 1949 they had already built *Tianlong Hao*, a sixty-horsepower fishing vessel which was regarded as the most advanced distant-water longliner in Kaohsiung. Four years later, they built *Siechang*, another longliner but with an eighty-horsepower engine. The two vessels fished in the waters off the Philippines and pioneered the revival of Kaohsiung's distant-water fisheries.[27]

The early postwar exploitation of the fishing grounds off the Philippines did not mean that the Pans were abandoning Taiwan's offshore fishing grounds. In the early 1950s, in order to fish for round herring (*Etrumeus terres*) and spotted chub mackerel (*scomber australasicus*), the Pans built three offshore vessels, *Tianfong*, *Tiansiang* and *Tian'ying*, which worked off Penghu and Suao, depending on the season. In the mid-1950s they built two sixty-horsepower vessels, *Tianhua Nos. 1 and 2*, for the offshore mullet fishery (*Mugil cephalus*). Every winter an enormous number of mullet migrated from northern China to the southern end of Taiwan to spawn before starting their homeward journey. The annual presence of valuable roe worked the offshore fishing communities into a state of "mullet frenzy." *Tianhua Nos. 1 and 2* moved northward to the waters off Tainan following the shoals of mullet, and then pursued them all the way to the southern tip of Taiwan before the fish started to migrate northward.[28]

The steady growth of the Pans' fishing business boosted their financial strength and encouraged them to diversify into shipbuilding. In the early 1950s they set up the Tian'er Shipbuilding Yard. This proved to be an astute business decision, for Kaohsiung's booming fishing industry brought an avalanche of orders for vessels. Within a few years Tian'er became one of the largest

---

[27]Pan San'guang, interview, 30 May 2002; and *The Present Situation of Taiwan's Fishing Industry* (Tokyo, 1978), 26. By 1954, Kaohsiung's longliners were already operating in the Celebes, Banda and Flores seas.

[28]Pan, interview, 30 May 2002.

private shipbuilders in Taiwan. In 1958 alone, 5.1 percent of the total tonnage of all newly-built vessels in Taiwan was constructed by Tian'er.[29]

In addition to the tremendous profits it generated, Tian'er brought an additional bonus to the Pans in that they were now able to build their own vessels without relying on other shipbuilders. Thus, the US aid they received for the construction of new vessels could go directly into their own pocket. The first vessel of this sort was built in 1956, and the steady receipt of US assistance became a standard means for the Pans to expand their fishing fleet.[30] Their first vessel constructed with US aid was *Yutai*, a 100-ton, wooden-hulled vessel with a diesel engine that was expected to exploit the tuna resources off Southeast Asia. It carried a crew of twenty, all of whom were Penghuans. With the most advanced fishing gear and a powerful engine, *Yutai* could reach the coasts of Singapore, Sumatra and even into the eastern Indian Ocean. Most of its catch was yellowfin tuna (*Thunnus albacares*), spearfish and shark. Such voyages took two to three months because no overseas supply bases had yet been established.[31]

In 1953, a modern tuna-canning factory was established by a US company, Van Camp Sea Food, in American Samoa, and longliners from Taiwan, Japan and Korea were invited to fish there.[32] *Siejin*, *Dongsheng* and several other Kaohsiung vessels were pioneers in Samoan waters, completing an unprecedented expedition in 1964 that represented a landmark in Taiwan's maritime development.[33] It was the first time that private Taiwanese vessels had undertaken such a long voyage into the southwest Pacific Ocean. Taiwan's

---

[29]*Ibid.*; and *Fisheries Tribune*, No. 17 (1962), 14.

[30]In 1956, thanks to US aid, a longliner-building programme was sponsored by the government in which ten distant water longliners were to be built. See *Fisheries Tribune*, No. 13 (1962), 5.

[31]Pan, interview, 30 May 2002.

[32]Jhang Sen, *The Management of Fisheries* (Taipei, 1977), 25. In 1963, another American company, Star Kist Foods Inc., also set up a canning factory in Samoa.

[33]Pan, interview, 15 May 2002; Sun Tai'an, *The Distant Water Longline Fishing of Taiwan* (Taipei, 1973), 8 and 26; and *The Present Situation of Taiwan's Tuna Longline Fishing* (Tokyo, 3). *Siejin* and *Dongsheng* belonged to Chen Cingbao, another well-known figure in Kaohsiung's fishing industry. As many as eleven Taiwanese private longliners first went to Samoa in 1964. Before that date, Taiwan's fishing companies knew little or nothing about the fishing grounds in the South Pacific. The first fishing company to use overseas supply bases and to fish in the Indian and Atlantic oceans was the China Fishing Company (CFC), but it was not successful. Much information on this may be found in the *Fisheries Tribune*. The CFC's efforts did not contribute to the establishment of Taiwanese global-scale fisheries due to constant mismanagement.

distant-water fishing industry had gained a major head start on other Asian fishing nations, except for Japan. Equally important, it was the first time that private Taiwanese companies had used overseas fishing ports as supply bases and cooperated directly with Western fish dealers. When the vessels returned to Kaohsiung a year later, residents gathered to greet the distant-water longliners and their crews. This initial success boosted the vessel owners' confidence in developing a global longline fishing industry. Some vessels now regularly voyaged toward Samoa, while others went westward and fished for albacore off Mauritius. In response to this emerging boom, the Pans also built *Taiyisiang* and *Taiyousiang*, 150-ton, wooden-hulled longliners with 450-horsepower engines. The two vessels made their maiden voyages to Mauritius in 1961. The success of the forward supply bases in Samoa and Mauritius inspired international fish dealers to establish additional bases in ports adjacent to major fishing grounds, such as Reunion, Cape Town, Abidjan and Las Palmas. The Pans also responded to the opening of distant fishing grounds by building 250-ton longliners. But a steel-hull was necessary for such large vessels, which posed a new technical challenge. To meet it, the Pans' engineers and technicians put considerable effort into collecting information about Japanese shipbuilding technology. Their work was in vain: *Fuyuan No. 1*, built in 1965, was one of the first Taiwanese civilian longliners to work the fishing grounds off Cape Town. Over the next few years, *Fuyuan No. 3*, *Taiyuan No. 1*, *Dayuan No. 11*, *Dayuan No. 12*, *Dayuan No. 21* and *Dayuan No. 22* were built and sent to fish in the Indian and Atlantic oceans.[34]

In the mid-1960s, the Taiwanese trawling grounds expanded into the southern part of the South China Sea. With a view to enhancing efficiency, a development programme for the distant-water pair-trawl fishery was launched by Taiwan's fisheries authorities. Five units of technically advanced pair-trawlers were built with government subsidies, and the Pans were allocated two units of 200-ton pair-trawlers: *Yongyuan Nos. 1 and 2* and *Yongyuan Nos. 11 and 12*. All four were equipped with state-of-the-art refrigerators that could maintain temperatures of thirty-two degrees below zero for months at a time, and they were expected by the government to be able to fish in the waters off Vietnam. With this new equipment, the Pans expanded their operations from Vietnam to Indonesia and eventually to the waters off northern Australia.[35]

A setback occurred in 1972 when *Yongyuan Nos. 11 and 12* were detained off Darwin for illegally fishing in Australian territorial waters. Pan

---

[34]Pan, interviews, 15 and 30 May 2002.

[35]*China Fisheries Monthly*, No. 375 (1984), 25-34; and Pan, interview, 30 May 2002. According to Pan, Taiwanese trawlers began to exploit Australian waters in the late 1960s - a bit earlier than the experimental vessel, *Haicing Hao*, which first arrived off northern Australia in 1971. Pan's recollections are verified in the article in the *China Fisheries Monthly*.

immediately flew from Kaohsiung, hired lawyers and sought the assistance of the Taiwanese Embassy to reclaim his vessels. Unfortunately, diplomatic relations between Taiwan and Australia were shaky at the time because the Australian Prime Minister, Gough Whitlam, was about to establish diplomatic relations with China, and all the Embassy staff was engaged in attempting to salvage the situation. The plight of Pan's two vessels was considered a low priority, and the Embassy offered no assistance. On his own initiative, Pan managed to retrieve one of his trawlers in the courts, but the other was confiscated and sold. The Pans were discouraged by this incident and did not build any further trawlers even after *Haicing Hao*'s success. Australia's subsequent implementation of a 200-mile EEZ forced the Pans sell all their distant-water trawlers and to concentrate on other fisheries.[36]

Chen Shengli was born in 1938 in Siao Liouciou, a tiny, impoverished island off the southern coast of Taiwan. The son of a fisherman, Chen began his fishing career at the age of thirteen. His competence and responsible attitude impressed his elders, and he was invited to share in the operation of a newly-built vessel while still in his late teens. That first vessel was a tiny ten-ton, six-horsepower longliner, and its only navigation device was a simple compass. The duration of this vessel's voyages never stretched beyond one week; in rough weather, especially in the winter, the vessel did not put to sea. Thus, in the early years Chen and his partners never operated beyond the Bashi Strait.[37]

The constraints of this tiny fishing boat frustrated Chen. As a result, he left his fellow crew to build his own longliners, *Baoyifa* and *Longjifa*. Their relatively powerful engines enabled them to operate in more distant waters. The eastern boundary of their fishing grounds was Lanyu, an island off the southeast coast of Taiwan, and the southernmost edge was the shoreline of Luzon. But a fishing career based in Siao Liouciou no longer satisfied Chen. The reconstruction of Kaohsiung Fishing Port had fostered the development of a wide variety of fishing enterprises and ancillary industries in the ten years after the end of World War II. Chen could not resist the port's attraction and in 1957 moved to Kaohsiung, where he ran a shop at Gushan Fishing Port selling rice, charcoal, fishing gear and groceries to fishers and fishing companies. At the same time, he raised funds to build his first longliner in Kaohsiung; this was *Lijincai*, a twenty-ton vessel with a twenty-five-horsepower *Yakitama* semi-diesel engine. It cost NT $160,000, which was more than enough to purchase a decent house in Gushan, one of the wealthiest areas in Kaohsiung at the time. The financial sponsorship of this vessel was shared by seven Siao Liouciou people, and Chen had a half-share in the vessel.

---

[36]Pan, interview, 30 May 2002.

[37]Chen Shengli, interview, 21 May 2002.

He now transformed himself into a vessel owner and never went to sea again. *Lijincai*'s fishing grounds were comparatively remote but extensive. Sometimes the vessel worked off Hong Kong in search of giant black marlin or in the waters off Philippines. During the typhoon season, it stayed in the safe areas of the offshore fishing grounds to take moonfish (*Mene maculate*), a small fish that could be used as bait.[38]

Chen was good at his business, and considerable revenues from his shop and *Lijincai* flowed into his coffers; in a short period of time, he earned an enviable reputation in Kaohsiung's Siao Liouciou fishing community. From 1961 onward, he created partnerships and expanded his offshore longliner fleet to include *Jiyuyu*, *Jinshengyu*, *Rueisheng*, *Manjinfa*, *Yumanzai* and *Rueicheng Hao*. Chen insisted that all the shareholders must come from Siao Liouciou (kinsmen and old friends were preferred) and that he had hold the largest share in each vessel so that he could manage this fleet. The tonnage of the vessels was still comparatively small, and the horsepower of the engines modest, so the fleet worked only in the waters off the Philippines. In March, the vessels regularly fished yellowfin tuna on the west coast of the islands, while in April and May they went to the east coast of Luzon to catch dolphin fish (*Coryphaena hippurus Linnaeus*).[39]

In those days, the marine resources on the longline fishing grounds off Southeast Asia were so abundant that voyages typically took only about ten days. This excellent fishing environment rapidly bolstered both his confidence and financial strength. He ordered more vessels in the early 1970s, including *Sinshengyi No. 1*, *Fashengyi*, *Shengjisiang*, *Singrueilong* and *Cyuansheng-siang*. Again, all the shareholders were from Siao Liouciou, and Chen held the largest share in each vessel. More powerful engines enabled these vessels to operate off Brunei at the southern end of the South China Sea. Traditionally, the Siao Liouciou people had no ambition to exploit remote grounds, so the longliners they built were much smaller than those of the local vessel owners in Kaohsiung; thirty or forty tons was usually the maximum. Chen, however, pioneered in building bigger longliners: *Fashengyi* was sixty tons, and *Singrueilong* and *Cyuanshengsiang* were both fifty tons. These longliners were considered distant-water vessels by the standards used by Taiwan's fisheries authorities.[40] Furthermore, *Singrueilong* and *Cyuanshengsiang* were both equipped with refrigerators that enabled them to undertake long voyages in distant waters.

---

[38]*Ibid*.

[39]*Ibid*.

[40]*Ibid*.; Chen Shengli, interview, 29 June 2002; Hong Fucai, interview, 5 January 2003; and *Survey on the Fishing Vessels and Fishing Gear of Taiwan* (Kaohsiung, 1989), 49-50.

His thriving business encouraged Chen in 1972 to build *Sinwunyi*, an eighty-ton longliner equipped with the best refrigeration available at the time. This new vessel was expected to operate in the waters beyond the South China Sea. Unfortunately, on its maiden voyage in 1973 *Sinwunyi* was detained due to an unlawful incursion into Indian territorial waters. The tremendous financial setback and worrisome legal problems caused by the detention of *Sinwunyi* were a huge blow to Chen. Hence, he decided to slow the further development of his longline business. From 1974 onward, Chen ran a gear shop that sold longlines, traces and other items and equipment for offshore fishing vessels, but this new venture only lasted three years.[41]

The establishment of important business connections between the Japanese fish dealers and the offshore fishers of Donggang generated an offshore longline boom in the fishing communities of southern Taiwan in the mid-1970s.[42] The profits from the offshore tuna fishery encouraged an exodus from Kaohsiung to Donggang. Some vessels from Siao Liouciou shifted their homeport to Donggang, while others decided to stay put. Chen, however, made an unusual decision. He and his onshore staff stayed in Kaohsiung, while his vessels oscillated between the two places (see figure 6.6).[43]

Figure 6.6: Chen Shengli and One of His Offshore Longliners Being Repaired in Drydock in Cianjhen Fishing Port

Source: Courtesy of Chen Shengli.

[41]Chen Shengli, interview, 8 July 2002.

[42]Cai Wun'yu, interview, 29 May 2002. A business relationship between Japanese fish traders and Taiwanese distant-water fishing companies had been established not long after the end of World War II; it was strengthened after overseas supply bases were established by Taiwan's fishing companies.

[43]Chen Hsien-Ming, "The Development of Tuna Longline Fishery on Liuchiu Island," *Geographical Research*, No. 33 (2000), 199. Some vessel owners moved to Donggang, while others did not. The reasons for this choice were sometimes very personal and had nothing to do with the fisheries.

Figure 6.7: Chen Shengli's Offshore Longliner, *Sinshengyi No. 2*, Loading Ice in Cianjhen Fishing Port

Source: See figure 6.6.

Figure 6.8: Chen Shengli's Offshore Longliner, *Jinlongfa*, Moored in Cianjhen Fishing Port, 1979

Source: See figure 6.6.

To cash in on Japanese demand was not as easy as Chen had thought because the quality of the fish required for Japan's *sashimi* market was high. The fish had to be perfectly fresh, and the skin had to be free of scrapes. Unfortunately, all his vessels were equipped with traditional Y*akitama* engines that produced constant vibrations at sea. As a result, most of his fish were bruised, and less than one-tenth of his catch met the stringent quality requirements of Japan's *sashimi* market. Chen soon learned that there was much room for improvement in the quality of his marine products. He therefore stopped running his fishing gear shop in 1976 and started to build a new series of vessels: *Sinshengyi No. 2* (see figure 6.7) *Yongchunyi*, *Sinlianyi*, *Jinlongfa* (see figure 6.8), *Sinyi*, *Sinshengyi No. 10*, *Jinyusheng*, *Sinshengyi No. 12* and *Jinrueiyi No. 1*. There were two major technical improvements in this new series of longliners. First, all of them were equipped with new *yama* (diesel) engines so damage to the fish was kept to a bare minimum. Second, to improve the operational efficiency of the vessels, the ratio of horsepower to tonnage was increased. This ratio traditionally was at least three to one, but *Sinshengyi No. 12* and *Jinrueiyi No. 1* had ratios of more than eight to one, and *Sinshengyi No. 10* and *Jinyusheng* were nearly eleven to one. The state-of-the-art equipment on these longliners and the power of their engines could cope with the most severe weather conditions, so typhoons, giant waves or communication failures were not considered nearly as threatening as before.[44]

The case studies of the histories of these Taiwanese fishing companies suggest several things. First, increased demand for marine resources made the most significant contribution to the expansion of the postwar trawl-fishing grounds. Over-fishing on the northern grounds in the 1950s drove a large number of Keelung's trawlers to relocate in Kaohsiung to fish in the South China Sea. Subsequent over-exploitation of these waters in the late 1960s forced Taiwanese trawlers to migrate south towards the waters between Indonesia and northern Australia. Private trawler owners were sensitive to changes in the marine ecosystem and market-driven demand. Numerous privately-owned trawlers moved to Kaohsiung to exploit the rich marine resources in the South China Sea once the problem of over-fishing emerged in the East China Sea. The staff numbers in the private enterprises were always kept to a bare minimum. A group of five or six well-trained onshore staff could efficiently control fleets operating in distant waters. When challenges arose, they could dispense with bureaucratic formalities and resolve them quickly. These small, private companies eventually became the main driving force of the emergent overseas Taiwanese fishing industry.

Second, there was a trend in the late 1950s for migrant fishers from various parts of Taiwan to use Kaohsiung as their homeport for a wide range of offshore fishing and related activities. A large number of trawlers from Keelung and numerous fishers from Siao Liouciou relocated to Kaohsiung to take advantage of its modern port facilities and to exploit the nearby marine resources off Southeast Asia. These developments confirm one of the main conclusions of chapter three that the excellent port facilities in Kaohsiung made it one of the most modern, convenient fishing harbours in East Asia, and that this established a solid foundation for the growth of Taiwan's postwar offshore

[44]Chen, interview, 29 June 2002; and *The Fishing Industry of Taiwan* (Nantou, 1953), 31. *Yakitama* engines (semi-diesels) produced constant vibrations at sea.

fishing industry. Third, trawler and longliner fleets both operated in Southeast Asian waters, but the scale of the expansion and its directions were different. Since ecosystems were hard to restore once they were damaged, trawler fleets always had to search for new fishing grounds. They left the East China Sea, moved along the continental shelf to the Taiwan Strait and the South China Sea and, in the early 1970s, finally arrived in the waters off Indonesia and northern Australia. In general, the expansion of Taiwanese trawl-fishing grounds was southward. By contrast, longline fishing was not as detrimental to the marine environment. While longliners operated by local tycoons soon left the waters off Southeast Asia for the Indian Ocean and the southwest Pacific in the 1960s, a new wave of offshore longliners, owned mainly by Siao Liouciou immigrants and local Kaohsiung fishers, quickly filled the vacuum.[45] In general, the longliner fleets working off Southeast Asia expanded their activities in a fan-shaped pattern; the eastern boundary of their grounds was the Philippines, and the westernmost edge was situated opposite Hainan Island (see figure 6.9). The selection of longline fishing grounds depended on seasonal conditions and/or what kind of fish they wanted to catch to meet current market demand.

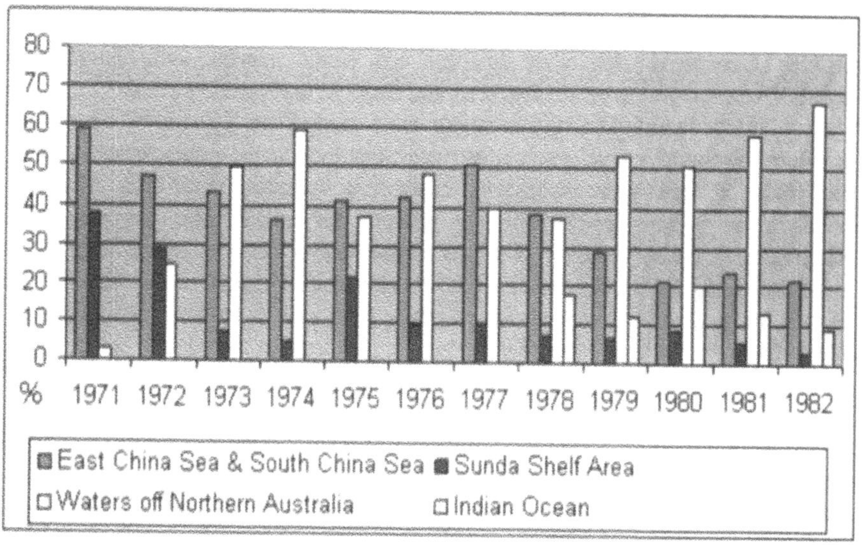

Figure 6.9: Proportional Change in the Catch of Taiwan's Distant-Water Pair-Trawl Industry in the East and South China Seas, the Sunda Shelf, Northern Australia and the Indian Ocean, 1971-1982

Source: See figure 6.2.

[45]Many offshore longliners still work in the Taiwan Strait and South China Sea.

Fourth, improvements in boat-building techniques and fisheries technology played a crucial role in the exploitation of new offshore fishing grounds. Old, inefficient *Yakitama* engines were replaced by new diesels that produced greater horsepower and fewer vibrations, thus reducing the likelihood of damage to the catch. Wood was replaced by steel as the main building material of the hull and superstructure, which made the building of larger trawlers possible. The installation of large refrigerators could keep the catch fresh on the long return voyage from northern Australia. Larger vessels also had bigger holds that enabled fishers to catch greater quantities on more remote fishing grounds, which also reduced operating costs. In addition, the higher ratio of horsepower to tonnage helped the new trawlers exploit remote fishing grounds more efficiently.[46] All these innovations in boat building greatly contributed to the development of the postwar fishing industry.

Fifth, the type of fishing method was strongly influenced by ethnic factors. Most local vessel owners re-entered fishing careers by operating longliners, but quite a few extended their interests into ancillary industries or trawl fishing. When the 200-mile EEZs came into force in the late 1970s, most who did not participate in international fisheries cooperation agreements either returned to distant-water longline fishing or entered gill-net or purse-seine fishing. Compared with other business and ethnic groups, local Taiwanese vessel owners were more adventurous and flexible during this period.

Most mainland Chinese began their careers on Taiwan by operating trawlers, although a few, especially from Shandong, tried other fisheries. But the scale of their new enterprises could not compare with their trawling endeavours. When trawl fishing's golden era ended, the majority left Kaohsiung's fishing circles. Yet this conservative attitude meant that their financial strength was not destroyed by the dramatic changes of the late 1970s. Hence, most Shandong vessel owners have been able to remain prosperous to the present.[47] On the other hand, their conservatism prevented them from participating in the global expansion of Taiwan's fisheries in the 1980s.

The most conservative group in Kaohsiung's fishing industry was comprised of people from Siao Liouciou, most of whom stuck to offshore longline fishing. They preferred to build a fleet of medium-sized offshore longliners rather than huge distant-water vessels. They worked the old fishing zones that were vacated by locals and expanded southward. They were not keen to explore the possibilities of distant fishing zones by themselves, nor did they like working with other ethnic groups, which further limited their potential for engaging in the fisheries expansion in postwar Taiwan.

---

[46]Lin, interviews, 23 May and 18 June 2002; and *Survey on the Fishing Vessels and Fishing Gear*, 114. The ratio of HP to T of the Taiwanese pair-trawlers in 1989 can be presented by the following formula: $HP = -128.36 + 3.38\,T$.

[47]Yu Jyun'yue, interview, 9 April 2002.

Chapter 7

Daily Lives at Sea, Fishing Zones and Politics

Since fishermen spent most of their time working at sea, their daily lives during fishing trips is an important, if neglected topic in Taiwan's maritime history. In this chapter I will discuss the fishers' roles, responsibilities and onboard politics. The fishing master, radio operator and engineman were the three most powerful figures on a vessel; I will explain the basis of their power and examine their interactions with other crew members as well as their personal feelings about their time at sea. Since the various fishing sectors had different practices which affected the daily lives of the fishers, I will map the fishing grounds as well as the major species targeted and their seasons.

The Fishers' Roles and Onboard Politics

Most of the time, the fishing master was the most experienced fisher. He might have come from a fishing family, or he might have started his career in his teens. He was apt to be semi-literate and to speak and write roughly. Nonetheless, he was extremely familiar with the ecology of his fishing ground and the range of technologies and techniques. In the 1950s and 1960s, the only navigation aids fishers had were nautical charts, compasses and sextants. A fishing master therefore had to rely on his experience and deep ecological knowledge which took years to accumulate.[1]

The fishing master was also the most powerful figure. In many cases, he was also a shareholder, so the owner had to respect him. Moreover, the he was in charge of the recruitment and employment of new fishers with the exception of radio operators. Those who wanted work had to curry favour with him rather than with the vessel owner. Some fishing masters even expected extra money from novices at the end of a voyage. While new recruits resented these practices, none expressed their feelings publicly because the fishing master also assessed their work performances in reports to his company. This made the fishing master so powerful that no one, except perhaps the radioman, would challenge his authority.[2]

---

[1]Chen Youyi, interview, 26 June 2002.

[2]Jheng Sanbian, interview, 20 June 2002; and Syu Yisin, interview, 4 July 2002.

On board the vessel, the fishing master was the "financial hub." He looked after the money the owner entrusted to him and took charge of all onboard financial affairs, especially the allocation of food allowances. Fishing masters could easily embezzle some of the money. Indeed, some fishing masters were like despots: anyone who doubted their honesty and leadership was sacked as soon as they returned to Kaohsiung. Most fishers chose to keep quiet even if the behaviour of their fishing master was incorrect.[3]

If necessary, crew members could borrow money from the fishing master. In addition to keeping a deck log, he had to maintain a small notebook that listed every financial transaction both on the vessel and in foreign ports. By virtue of his status, the fishing master was financially privileged. In foreign ports the crew had to pay for their own entertainment, but a fishing master was allowed to claim an allowance for this from his company. Hence, it was common for him to inflate his expenses. The vessel owner did not usually question such fraudulent behaviour as long as the fishing results were good.[4]

Fishing masters served as intermediaries between the fishing companies and the fishers. To prevent their authority from being challenged, some fishing masters chose to remain aloof, appearing only to give orders. First mates served as his intermediary. In the eyes of some fishers, the first mate was the fishing master's "flunky." Although physical punishment was strictly prohibited, it was still frequent. Some fishing masters personally beat crew members for insubordination or dereliction of duty, while others left this to their first mates. As Syu Yicin, one of my informants, said, "[t]here was no law at sea, only violence."[5]

Both the fishing master and the crew could earn large sums on each trip, but in the early years many came to a bad end. After a long, tiring and abstinent voyage, quite a number spent money lavishly on rest and recreation. Some visited prostitutes to satisfy their sexual desires; some indulged in heavy drinking; and some gambled until they ran out of cash.[6]

Yet being the fishing master on an offshore longliner run by locals from Kaohsiung was a totally different story. First, he did not have to bother to recruit because all his men were relatives, friends or fellow villagers, so

---

[3]Chen Youyi, interview.

[4]*Ibid.*

[5]Syu, interview.

[6]Chen Youyi, interview; Sia Fujhong, interview, 24 April 2004; and Liou Syuan, interview, 4 July 2002.

those who wanted work did not need to curry favour with him.[7] Second, he was not as personally powerful as captains on distant-water vessels or Siao Liouciou longliners. When he talked to his crew, he needed to speak respectfully, because some fishers might be older members of his family. Under such circumstances, he needed to consult them whenever there was a problem. Third, being the fishing master of a tiny vessel meant that the master did not really earn much more money than the other fishers. They went to sea together, and when they returned from their fishing, they sold the catch at a port market together; therefore, sharing the profits from the catch was completely transparent.[8]

The only man at sea who dared challenge the fishing master was the radio operator. His expertise in telecommunications and his relationship with the military gave him confidence to deal with the fishing master on an equal footing. The training and allocation of radiomen was strictly controlled by the military. The slow process of training could not meet the industry's growing demand in the postwar years. Hence, there was always a shortage of radiomen. This was highlighted in 1964 when Taiwanese distant-water longliners were invited to exploit the fishing grounds off Samoa. To compete effectively for wireless operators, a "guaranteed wage system" [*Baojheng Daiyu*] for the radiomen was introduced by Kaohsiung's fishing companies.[9]

In many cases, a fishing master did not get on with his radioman because the latter was responsible not only to the fishing company but also to the Taiwan Garrison Command. Indeed, radiomen were obliged to collect information for the military and to notify the government of their findings. Often there was a conflict of interest between the radioman and other crew members. To maximize their fishing results and hence their profits, some fishing masters took the risk of fishing in "Prohibited Zones." In such circumstances, should a radioman inform the government? If he did so, his

---

[7]M. Estellie Smith, *Those Who Live from the Sea: A Study in Maritime Anthropology* (New York, 1977), 28-29. The fishing master chose his own crew from among his relatives or people from the same or nearby villages; such people can easily be found in the fishing communities engaged in offshore fishing. In the fishing communities on the Scottish east coast, the fishing master also recruited fishers from "a small circle of relatives, townsmen, or fishers from nearby villages."

[8]Cai Bian, interview, 20 April 2002.

[9]Jhong Ciourong, interview, 15 April 2002. Work as a radioman was considered an enviable job in Taiwan's fishing industry because such men earned decent money without doing any heavy work. The same is found in the British fishing industry. A radioman in Hull did not need to do "a physically demanding job" but just operated his telecommunications equipment on the bridge and earned good money. The deckhands were upset because the radioman earned as much as they did without ever touching a fish. See Jeremy Tunstall, *The Fishermen* (London, 1962), 129.

fishing master and crew would be in serious trouble. Furthermore, fishing masters were usually suspicious of the radioman because they were wary that he might report something negative to the vessel owner, especially if the catch was poor. Besides, the radioman was often the most educated person on the vessel and he sometimes questioned the instructions of the fishing master in front of other crew, something that at the very least made the master uneasy.[10]

Conflicts between the fishing master and the radioman never happened when the latter also was qualified as a fishing master. Being a fishing master did not necessarily mean that he could retain his position forever. Following two or three unsuccessful fishing trips, he often was assigned to work as an ordinary fisher. Some fishing companies hired a less competent fishing master as the "nominal fishing master" and then requested a competent fisher to do much of the work of a real fishing master. In many cases, the "real fishing master" was a radioman. The first radioman to work concurrently as a fishing master in Kaohsiung's fishing industry did so about 1966. Fishing companies began to adopt this approach not only because the radioman was often the most educated person but also because in this way they could quickly learn new fishing techniques and acquire the nautical knowledge required for the distant-water fisheries.[11] This system could also reduce the wage bill considerably. A case in point was the bonus-sharing system. A fishing master might receive seven to ten bonus shares, while a radioman and engineman received just three each. If a vessel owner hired a fishing master and a radioman, he would need to provide ten to thirteen bonus shares between the two. But if an owner appointed his radioman as the fishing master, he only needed to pay between four and seven more shares to the radioman and another half a bonus share to the "nominal fishing master." Hence, the company saved the equivalent of nearly three bonus shares each trip.[12]

Usually, all parties involved were happy with this system. The nominal fishing master could earn an additional half bonus share simply because he "offered" the use of his fishing master's licence. Hence, the radioman could work informally work as the real fishing master under this scheme without a captain's licence. Yet neither the military nor the fisheries authorities were comfortable with the system because it made it difficult to monitor fishing activities properly. Although this dual staffing system was declared illegal, the practice still existed in Kaohsiung in the 1960s and

---

[10]Jhong, interview, 15 April 2002; and *Fisheries Tribune*, No. 97 (1971), 12. Fishing companies and fishing masters could not control radiomen well.

[11]Jhong, interview, 15 April 2002.

[12]*Ibid.*

1970s.[13] There are two other things to emphasize here with respect to this staffing technique. First, a radioman never worked concurrently as a fishing master on offshore longliners run by Kaohsiung locals or those from Siao Liouciou because most of their vessels were less than fifty gross tons and hence did not require a radioman. Second, radiomen were gradually phased out in the late 1970s due to the advent of new telecommunications equipment.

The chief engineman was the onboard mechanic. His duty was to look after the engine to ensure that it worked properly at sea. He also sometimes helped to refrigerate the catch. Although he was an experienced fisher, he was not directly involved with the actual activity of fishing. He had many people at his disposal to assist him. For example, in a vessel that carried twenty-two fishers, six crew usually worked in the engine room. Besides the chief engineman and his deputy, there were three or four assistants. The deputy engineman and the assistants had to stay in the engine room to monitor engine functions. Since the engine room was intolerably hot and noisy, the staff had to work in shifts. Apart from their engine room chores, they also needed to help the fishers unhook fish on deck. The time spent between these two tasks was divided roughly fifty-fifty. Most engineman assistants were new hands or had recently been promoted from the position of cook. Working as an engineman's assistant was in fact not significantly better than toiling as a cook.[14]

Things were different on board offshore longliners run by Kaohsiung locals because their vessels were too small to have an engineman's assistant. Moreover, often the engineman was the fishing master and had to look after the engine and the fishing. This not only saved on manpower but also made fishing operations more efficient.[15]

Some new recruits started at the bottom as a cook's boy. Most remained in this humble position for at least two years until they were strong enough to cope with the demanding deck tasks. Being a cook could be a miserable job; indeed, it often was the busiest post on a distant-water vessel. Most cooks were teenagers who had recently graduated from primary school. As a cook, they needed to do all sorts of kitchen chores as well as learn fishing skills from their seniors. The cook usually was asked to guard the vessel while in port. Apart from those additional tasks, he was often a target for the other fishers' anger and frustration; everyone on board tended to bully him. Although cooks were fishers, the crew did not treat them as such.[16]

---

[13]*Ibid.*

[14]Chen Youyi, interview.

[15]Cai, interview.

[16]Jheng, interview, 13 March 2002 and 20 June 2002; and Chen Youyi, interview.

A cooks on a pair-trawler, however, was never considered to be a fisher. He could be either a young adult or an old man, but he was rarely a teenager. He was also respected on the vessel. In some cases, trawler cooks had already worked as cooks ashore before going to sea. As a result, most had excellent cooking skills and were kept busy at this task from dawn to dusk. If a trawler pulled up its net four times per day, the cook would need to prepare four meals a day. In principle, the cook on a trawler was not permitted to get directly involved in fishing operations, although some did.[17]

On offshore longliners run by Kaohsiung locals, the oldest fisher usually filled the post of cook since his physical strength could not compare with that of younger men. Working on a tiny boat, the cook could not prepare meals when the weather was bad and the waves high. Indeed, no one expected him to cook under such circumstances. Instead, they expected him to suspend his cooking and help other crew members cope with the rough weather.[18]

When fishers were in Kaohsiung awaiting their next trip, their companies did not offer any financial aid. Fishers ashore needed to arrange their own accommodation and meals and thus often needed advances from the companies. Some of this money was used to buy the daily necessities they used onboard, and the rest was sent home to their families. While docked, the youngest fisher was required to stay onboard to guard the vessel. This was a form of exploitation because he was not paid for the task.[19]

Fishers had to endure many personal hardships and dangers during the course of their voyages. Some vessels did not have proper bathrooms, and fishers had to wash with sea water on deck. Moreover, because some vessels did not have toilets, fishers had to urinate and even defecate over the deck rail. Sometimes they fell into the water and drowned while relieving themselves. This most tragic occurrence often happened at night.[20]

Fresh water was life to fishers. Since they always sweated, they were allowed to drink as much water as they wanted. At the beginning of a voyage, they ate vegetables. After the vegetables and fruit were consumed, they could only eat preserved meat or fish. The food that fishers ate on longliners was

---

[17]Jhong, interview, 15 April 2002. Since the cook was not considered a fisher, the lowest status fisher on board a trawler was the assistant engineman. See Tunstall, *Fishermen*, 119. On British trawlers the social status of cooks was comparatively higher than other crew members, but this does not mean that they received higher wages. This hierarchical system is similar to Taiwan's trawling industry.

[18]Cai, interview.

[19]Chen Youyi, interview.

[20]Syu, interview; and Cai, interview.

extremely basic: rice, fish, soup and sometimes vegetables. Salt and oil were absolutely essential for cooking and nutrition, and any shortage of these items could virtually paralyze operations. Fishers had fish at every meal as the main source of protein.[21] Meals on trawlers were comparatively better than on other fishing vessels since they returned to Kaohsiung to be re-supplied every month or two; hence, their food was fresher than on distant-water longliners. Offshore longliners also came back to Kaohsiung often, but the cook could not prepare food in rough weather due to the size limitations of the vessels.

Fishers rarely got enough sleep while at sea. Working sixteen hours a day on a distant-water vessel was considered normal. Sometimes the men were awake for more than a day when they were really busy. In extreme cases, longliner fishers worked up to eighteen hours at a stretch and only had four hours sleep per day. This state of affairs could last for nearly a month. Furthermore, berths on these vessels were extremely small. On a tiny offshore longliner, all fishers, including the master, had to sleep together in the hold. In the summertime, when the weather was warm and the sea calm, they often slept on deck. On a distant-water fishing vessel, the fishing master, engineman and radioman had their own cabins. The allocation of personal space accurately reflected the stratification of status and power among fishers. Since the fishing master was in charge of the voyage, he was not supposed to mingle with other fishers except when giving orders from the bridge. As the principal onboard mechanic, the engineman had several assistants to carry out his orders. The radioman needed to look after the code books and all sorts of classified materials; hence he also needed a room for privacy.[22]

Besides fishing, what kind of personal entertainment did they have during their long trips? They certainly gambled if they were allowed to do so. Indeed, some fishers were addicted to gambling; at home they gambled with their families and friends, often they gambled in foreign ports and sometimes they gambled at sea during breaks. Gambling in ports was a fisher's personal business; fishing masters usually would not intervene. But gambling at sea required the fishing master's permission. On some vessels it was strictly forbidden because masters believed that gambling would bring bad luck that might cause poor fishing or an unexpected mishap.[23]

Many fishers also sought spiritual comfort and security at sea through worship. Most fishing vessels carried statues of a religious figure as their patron saint. What particular gods fishers worshipped onboard was decided on their behalf by the fishing company or the fishing master. Traditionally, vessel

---

[21]Syu, interview; Jheng, interview, 13 March 2002; and Chen Youyi, interview.

[22]Syu, interview.

[23]Jhong, interview, 15 April 2002.

owners dared not discourage animist beliefs, even though they might be Christians.[24] The gods the fishers usually worshipped came from either the pantheon of Taoism or Buddhism. Still, it seems that fishers preferred Taoist gods rather than Buddhist *boddhisatvas* because they killed fish every day, which violated the basic precepts of Buddhism. In addition to patron saints, fishers worshipped the Heavens and the gangway of their vessel. They worshipped the Heavens on the first and fifteenth day of the lunar month, a tradition that many Taiwanese families practiced ashore. As for the gangway, it was the gate through which the longlines and fish entered and left the vessel, and fishers believed that they would catch more fish by worshipping at the vessel's gate. This "gate ceremony" was held twice: once before they went to sea and again when they arrived at their destination before starting to fish.[25]

Living in a confined space, Syu Yisin suggested, made fishers easily agitated, sometimes hot-tempered and occasionally brutal. When they communicated with each other at sea they tended to shout. Criticism and scolding occurred almost every day, and sometimes the men even cam to blows. Murders and "accidents" sometimes occurred. Any tool on the vessel, such as a knife, stick, cooking cleaver or steel rope, could become a weapon. Working on a fishing vessel to some degree was like suffering in a "water dungeon." Some fishers had a strong sense of being exploited and treated inhumanely.[26] Some, mainly on large distant-water vessels, were sometimes forced to form factions for protection. This happened partly because crew members did not necessarily know each other well before they put to sea and partly because they became

---

[24]Arne Kalland and Brian Moeran, *Japanese Whaling, End of an Era*? (London, 1992), 165. The tradition of fishing companies sponsoring religious ceremonies is also found in Japan. Taiwanese vessel owners go to the shipyard for a religious ceremony to bless a new boat. Japanese management also does the same thing, and fishing companies in both countries always put a guardian deity in each of their boats. On Taiwanese vessels, the guardian deity was often placed in the fishing master's room; on Japanese whalers, it was often placed below the harpoon gun.

[25]Syu, interview. See also Kalland and Moeran, *Japanese Whaling*, 166. A similar practice can be found in Japan's whaling industry. Japanese whalers gather on deck and hold a ceremony before the boat leaves for the Antarctic. When they hunt whales, like Taiwanese longlinermen they perform a number of rituals in front of the altar in the hope of having good results.

[26]Syu, interview. Horrible onboard working conditions can also be found in the fishing industries of some Western nations. In the UK, for example, the distant-water fisheries "placed great pressure not only on fish stocks but also on the men that fished them." In the postwar era, many British fishers still worked "in conditions that few modern landsmen could or would tolerate." Their situation was similar to that of Taiwanese fishers. See Robb Robinson, *Trawling: The Rise and Fall of the British Trawl Fishery* (Exeter, 1998), 190.

increasingly irritable during long, difficult voyages. One of my older informants, a man who was nearly seventy-two-years-old when I interviewed him in 2002, was still full of sorrow and bitterness as he recalled how his seniors bullied him at sea.

Taiwanese Fishing Zones in Southeast Asia

In the early 1960s, the majority of Taiwanese pair-trawlers still operated on the East China Sea, Taiwan Strait and the northern part of the South China Sea. The waters off Hong Kong served as an alternative for pair-trawlers that worked the Taiwan Strait. Every year, to avoid heavy seas in the Strait caused by strong north winds, large numbers of Taiwanese trawlers went south to fish off "the feet of Hong Kong." Their main target species were threadfin breams and groupers.[27]

Seasonal variations or dangerous weather were seldom the deciding factors when fishers selected their fishing grounds. Instead, Taiwanese fishermen "fished wherever fish might be." The rapid depletion of stocks in the Taiwan Strait in the 1960s was the main reason fishers shifted to the South China Sea. Pair-trawlers from Kaohsiung, except those that were old and small, fished in Tourane, the Gulf of Tonkin and the waters south of Hainan from the 1960s. Some even fished south of Saigon. Fishing trips from Kaohsiung to Tourane took about four days, and from Kaohsiung to Saigon approximately six days. Their target species, mainly off Tourane, were yellowback sea bream, edible cuttlefish, silverfish and lizardfish. All these species were very popular in Taiwan's fish markets.[28]

The waters off Saigon was a virgin zone for Kaohsiung pair-trawlers in the 1960s. The marine ecosystem had not been damaged, and the resources were rich compared with the grounds off Hainan Island or in the Gulf of Tonkin. A large proportion of the catch off Saigon comprised lizardfish, *ayu* fish and some coarse varieties which were not considered valuable. The primary reason that fishers congregated there was for the schools of cuttlefish and filespot snapper. Filespot snapper was very popular in Japan, and it was one of the few species that Kaohsiung's trawling industry exported there. The economic value of a particular species was determined by consumers' tastes. For example, yellow sea bream was popular in Taiwan, but Japanese consumers never liked it. Most of the catch from Taiwanese pair-trawlers was sold in domestic markets. Only a small proportion was exported to overseas

---

[27] *China Fisheries Monthly*, No. 375 (1984), 25-34; and Jhong, interview, 9 April 2002.

[28] Syu, interview; Jhong, interview, 15 April 2002; and Li Jijhao, interview, 22 March 2002.

markets after being processed at plants in Kaohsiung. Those vessels that worked off Saigon or Tourane seldom put into port there. The fishing grounds off Vietnam were not that far away from Kaohsiung, and fishers did not find it necessary to get provisions and supplies there.[29]

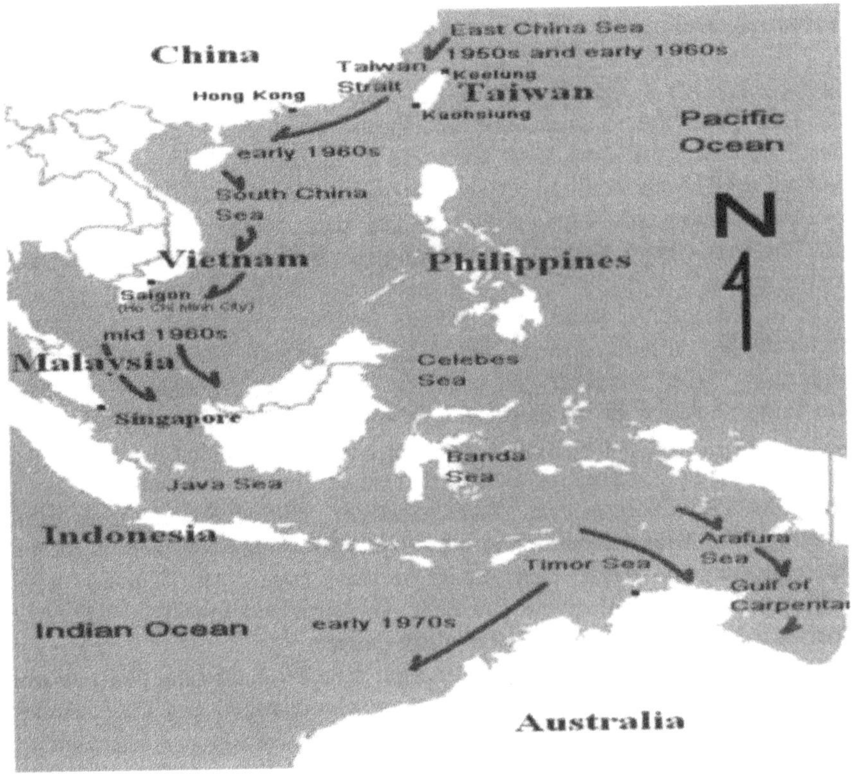

Figure 7.1: Expansion of the Taiwanese Distant-Water Trawl-Fishing Grounds

Source: Courtesy of the author.

The trawling grounds continued to expand southward, as vessel owners kept updating their vessels and onboard equipment (see figure 7.1). The widespread use of refrigerators enabled fishers to remain at sea for far longer periods. From the 1960s onward, a growing number of Taiwanese pair-trawlers ventured into the waters off the Sunda Shelf and North Borneo. Fishers called the grounds off the northern coast of Borneo "the head of Borneo." Meanwhile, the fishing grounds on the Sunda Shelf area were called "the Shoals" by elderly fishers. Marine resources there were rich, but the

---

[29]Jhong, interview, 15 April 2002.

small size of the fish did not really interest Taiwanese consumers. Due to the geographic proximity of the two fishing grounds, fishers could readily go to fish at the "feet of Saigon" and, if the catch was unsatisfactory, move to the "head of Borneo" and *vice versa*. Fishers continued this "to-and-fro fishing" for a number of years until the resources in both zones were totally depleted.

In the 1970s, Taiwanese trawlermen began to conduct private "fishing experiments" in Indonesian waters and the Indian Ocean, but not many fishing masters understood or recognized the territorial boundaries of other nations when fishing. Some captains were so bold as to fish within the territorial waters of coastal nations, which was viewed locally as a criminal activity. As a result, four or five vessels were confiscated by the Burmese government at the beginning of the 1970s for illegally fishing in its territorial waters. Confiscations also happened in Indian waters. As a result, crews tended to want to stay in the waters of Southeast Asia, which were deemed safer and were much closer to their home port.[30]

Nevertheless, the rich marine resources in the waters off Indonesia attracted a large number of Taiwanese fishers. Every year from August to December many Taiwanese pair-trawlers gathered in Indonesian waters, especially near the fishing grounds off Ambon Island, to catch Formosan squid, which could generate considerable revenue for both vessel owners and fishers. From January to July the pair-trawlers moved even further south to fish in the Gulf of Carpentaria. The "to-and-fro fishing" now shifted to the more distant waters between Indonesia and northern Australia. The Gulf of Carpentaria, especially the eastern section, was extremely rich in lobsters, Indo-Pacific king mackerels, silver sillago, narrow-barred Spanish mackerels and Formosan squid. All these species were popular in Taiwan's fish markets and restaurants, and silver sillago was also in demand overseas.[31]

Larger vessels with more powerful engines could travel faster than their predecessors. Therefore, even reaching the remote fishing grounds took less time. The overall length of a fishing trip to northern Australia, for example, was about fifty days, while the voyage from Kaohsiung to the inner part of the Gulf of Carpentaria took twelve days. Therefore, fishers usually worked about twenty-five days when off Australia. Most vessels fishing in this region had large fish holds and the best refrigerators, and they were fully laden with fuel. Maintaining ice and fuel supplies was no longer a problem, and the holds were usually full when they returned home. A 300-ton vessel operating off Australia could make four trips per year. But when the 200-mile Exclusive Economic Zone (EEZ) came into effect in Australia in 1978, the Taiwanese fleets were banned from the Gulf of Carpentaria. Fishers were forced back into

---

[30]*Ibid.*

[31]*Ibid.*

the waters off Indonesia and were not allowed to return to Australian waters until a cooperation agreement between the two nations was negotiated.[32]

In general, the exploitation of the longline fishing grounds off Southeast Asia in the postwar era occurred in two phases. The first spanned the years from 1945 to the beginning of the 1960s, while the second covered the years since then. As noted in chapter two, Kaohsiung's distant-water longline fishery was revived soon after the end of World War II. Some fishers from Kaohsiung started to fish in foreign waters again as early as 1946. The gross tonnage of their longliners ranged from twenty to 100 tons. Most vessels were owned and operated by companies in Gushan, which was the centre of the Taiwanese distant-water fishing industry before a modern fishing port was built in Cianjhen. People in the early postwar industry called their fishery a "distant-water fishery," which it obviously was compared to the colonial era.

The first fishing grounds to be re-exploited were off Luzon and the inland seas of the central Philippines that were referred to as "the Rivers" by Taiwanese fishers. They began their fishing trips from Kaohsiung, navigated along the western coast of Taiwan and then headed south through the Bashi Strait to northern Luzon. From there they navigated along the west coast of the Philippines to reach the inland seas. Their targets were shark and tuna. Although the vessels in the first years after World War II were still quite small, a vessel of about twenty tons still enabled fishers to work off the Philippines without undue difficulty. With the advent of larger longliners int eh 1950s, more Taiwanese fishers worked in the western part of the South China Sea. To get there, they had to traverse the northern part of the South China Sea and then navigate southward along the coast of Vietnam. In a short period, Taiwanese fishers also expanded into the waters off Indonesia and Malaysia. Fishing in the southern part of the South China Sea did not require big vessels. A forty-ton longliner with a crew of fourteen was capable of operating comfortably in this region. The length of such a fishing trip was approximately thirty to forty days; longline fishers could make about five such trips a year. Before 1963 most longliners still used Kaohsiung as their home port to fish in the waters off Southeast Asia and returned there when their fish holds were full. Old fishers called this oscillating fishing pattern "*dangpangha.*"[33]

They fished in a fan-shaped pattern across the region. The eastern boundary of their fishing grounds was the Philippines, the western edge was opposite the province of Hainan and the southern boundary was off Indonesia and Malaysia (see figure 7.2). Some larger longliners even went through the Malacca Strait in search of new fishing grounds. Before the 1960s, Taiwanese

---

[32]*Ibid.*

[33]Pan San'guang, interview, 15 May 2002; and Liou Syuan, interview, 17 March 2002.

fishing companies relied upon foreign ports as supply bases. The most distant fishing ground they could reach was off Penang on the eastern rim of the Indian Ocean. For example, *Sieshun* was a 100-ton, distant-water longliner capable of long-range fishing in the heart of the Pacific or Indian oceans. Yet it only operated off Malaysia and Indonesia, returning to Kaohsiung before its ice melted. Using such a large vessel to operate on neighbouring fishing grounds wasted its potential. The situation did not improve until Taiwanese fishing companies began to cooperate with international fish dealers and adopt the supply base system in 1963. Therafter, those fishers who operated on the longline fishing grounds off Southeast Asia became pioneers in exploring and exploiting worldwide fishing grounds. Generally speaking, they targeted two different fishing zones. Those who fished in the South Pacific left Kaohsiung directly for Samoa, while those who wanted to fish in either the Indian or Atlantic oceans (or perhaps even the Mediterranean) had to pass through the Malacca Strait. The vessels that used Samoa as a supply base were small compared to those that operated in the Indian and Atlantic oceans.[34]

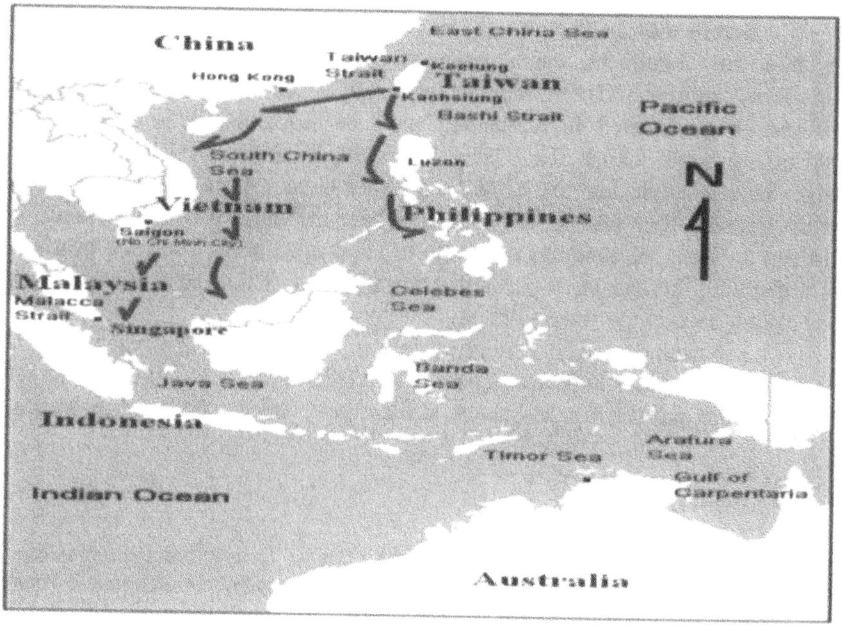

Figure 7.2: Fishing Grounds of the Kaohsiung Distant-Water Longliners before the 1960s

Source: See figure 7.1.

[34]Chen Youyi, interview; Jheng, interview, 20 June 2002; Pan, interview; and Huang Sianchih, interview, 6 May 2002.

It is interesting to note that in the colloquial languages of Kaohsiung fishers, "fishing abroad" means using foreign fishing ports as forward supply bases. In other words, fishing in the waters off Southeast Asia was still considered "fishing in the country." Singapore and/or the Malacca Strait were viewed as the gateways to distant "foreign" fishing grounds. When the distant-water longliners left the seas off Southeast Asia and ventured into the Indian and southwest Pacific oceans in the 1960s, a new wave of offshore longliners, mainly owned by Siao Liouciou migrants and local Kaohsiung fishers, quickly filled the vacuum on the grounds off Southeast Asia.[35] Compared with the fishing activities during the initial phase, offshore longline fishing in Southeast Asian waters in the 1960s had three basic features. First, most longliners were family-run. Vessel owners and stakeholders were relatives or neighbours from the same fishing village, and sometimes they fished together on the vessel they shared. Second, their vessels were much smaller; hence they had to expand their fishing grounds gradually. Third, they simply called their new fishing pattern the "offshore longline fishery," although some of the vessels were in fact capable of working off Indonesia and Malaysia in the 1970s.[36]

Before the 1960s, the only modern navigation tool these fishers had on their tiny longliners was a compass. Nautical charts and radios were considered luxuries. If the weather was good, radio reception and hence wireless programmes from Kaohsiung could be heard clearly in the northern part of the South China Sea. Using the wireless, experienced fishers could judge their location and the distance to Kaohsiung. When they began their return journey, they just set a course to where the radio waves emanated. In addition to using the compass and radio to navigate, fishers also took bearings by identifying the terrain of particular coastal areas. This traditional approach was mainly used in the waters off the Philippines.[37]

Kaohsiung's offshore longline fishers also expanded the reach of their more distant activities in a fan-shaped pattern (see figure 7.3). Some fished off Hainan Island and the Pratas Islands, while others fished off the western coast

---

[35]See chapter 6.

[36]Takahashi Jun-ichi, *Japanese Whale Culture: Tracing the Development of Whale-hunting Culture* (Kyoto, 1992), 132-134; Kalland and Moeran, *Japanese Whaling*, 1 and 5; and Jhong, interview, 15 April 2002. The phenomenon of offshore and distant-water longline fishing coexisting is not unique. In Japan, offshore whaling enterprises also coexist with distant-water whaling. STCW fleets (small-type coastal whaling) hunt whales in Japanese coastal waters, while LTCW fleets (large-type coastal whaling) work in the waters off Antarctica. Each has its own fishing grounds, markets and operational modes. Their relation is similar to that between offshore and distant-water longline fishing in Kaohsiung.

[37]Hong Fucai, interview, 21 May 2002; and Cai, interview.

of Luzon. Their vessels were still small; most were less than twenty tons and could only carry a crew of six or seven. It took about three days to reach their target areas. The ice in their vessels could last for only twelve days. In other words, due to technological constraints they could only fish for five days, which was not very efficient. During their outward trips, they needed to catch moonfish to use as bait. Some fishers could obtain sufficient baitfish for the duration of the voyage in a single day, while others had to spend four or five days to meet their requirements. If the baitfish catch was not satisfactory, the fishers were forced to return to Kaohsiung for more ice. The basic cost of fuel, ice and daily necessities was a considerable financial burden for these smaller offshore longliners. Going back to Kaohsiung with a small catch not only increased costs but also placed a heavy mental burden on the fishers with respect to their next fishing trip.[38]

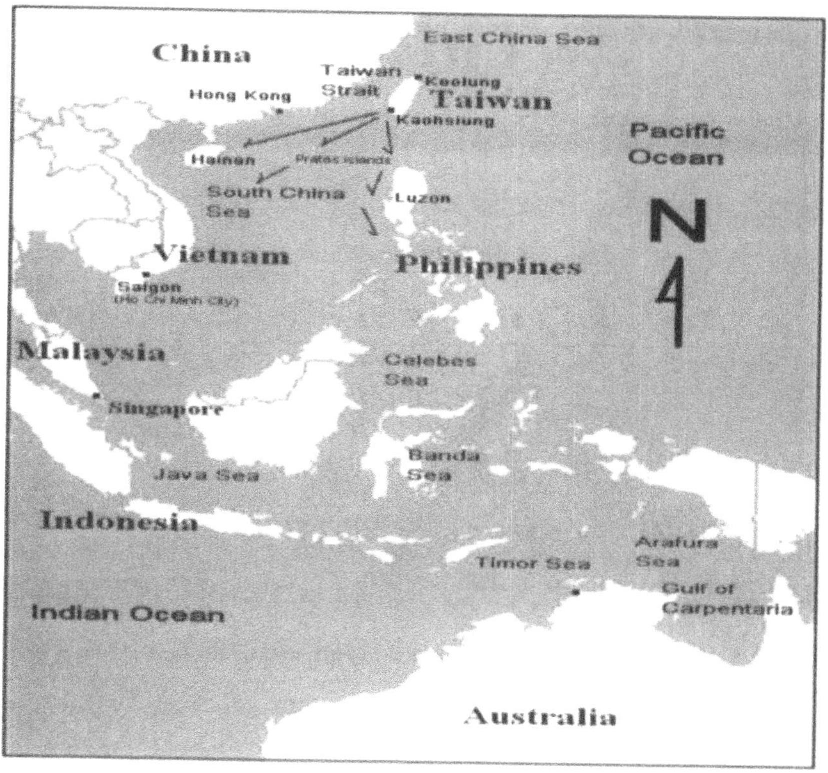

Figure 7.3: Fishing Zones of the Kaohsiung Local Offshore Longline Fishery

Source: See figure 7.1.

[38]Cai, interview.

Longline fishers went to certain grounds to fish for specific species. The western half of the South China Sea, the waters off Hainan Island and the coastal waters off Vietnam were considered shallow seas with plentiful stocks of black marlin (*Makaira indica*). The eastern half of the South China Sea and the waters off the west coast of the Philippines were considered deep-water areas with an abundance of yellowfin tuna (*Thunnus albacares*). According to the fishers' understanding of the marine environment, the Pratas Islands were the physical boundary between the shallow and deep waters.[39] In the summer season between June and August, fishers preferred to work in the shallow waters off Hainan Island. In the winter they would shift to the waters off the Philippines in search of deepwater fish like tuna.[40]

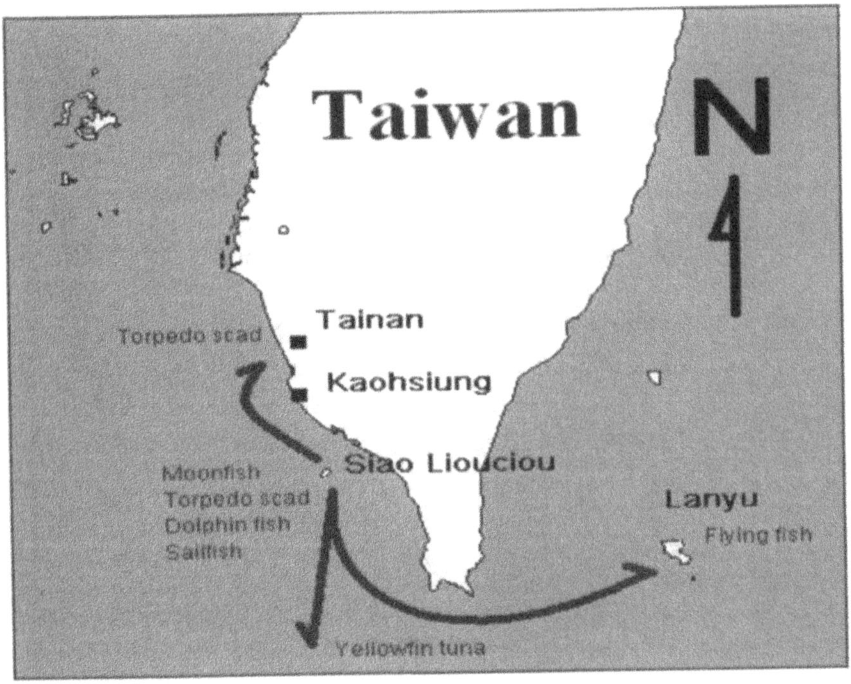

Figure 7.4: Fishing Grounds of Siao Liouciou Fishers before Moving to Kaohsiung

Source: See figure 7.1.

---

[39]This was the old fishers' understanding of the major differences in sea depth and species variation. But their views were shared by many others as well.

[40]Cai, interview.

Before migrating to Kaohsiung in the 1940s, people from Siao Lioucciou fished only off the "feet of Siao Lioucciou." The targets were dolphin fish and sailfish, and the limited fishing season ran only from April to June. Their vessels were tiny (about ten tons), and their fishing trips never lasted more than a week. These fishers had to stay ashore in rough weather because their vessels were so small. But those few with larger vessels tried both to explore and exploit more remote grounds. Some went in search of flying fish in the waters off Lanyu, a tiny island off the eastern coast of Taiwan, while others went south to hunt for yellowfin tuna (see figure 7.4).[41]

In the 1950s and 1960s, a growing number of Siao Lioucciou fishers moved to Kaohsiung. Armed with better port facilities, they started to operate larger longliners just as their Kaohsiung counterparts did. Some fished off Hong Kong, while others ventured into the waters off the Philippines. But no matter where Taiwanese fishers went, they first had to catch bait, which was kept alive in a special tank below deck. These fishers normally did not start to fish until they had enough bait. Like the Kaohsiung locals, if they were lucky it would only take one to two days to get this. But if they were unlucky, the process could last as long as a week, which would invariably make their trip unprofitable. The kind of baitfish Siao Lioucciou fishers used was determined by the season and the weather. Before migrating to Kaohsiung, Siao Lioucciou fishers had to stay in coastal waters to avoid rough weather in the winter months, fishing for offshore species like moonfish or torpedo scad. After winter ended, they would fish for squid or head north to the waters off Tainan County in search of torpedo scad.[42]

The local fishing grounds were not abandoned by Siao Lioucciou fishers after they migrated to Kaohsiung. No matter which fishing grounds they were going to work, they would still have to go back to obtain baitfish off the Siao Lioucciou coast. This was always a pleasant task because the vessel owners allowed fishers to go ashore at Siao Lioucciou for a day to visit their relatives and friends. Also, returning to Siao Lioucciou did not necessarily take any time away from the voyage because it was directly on their way to the fishing grounds off Southeast Asia.[43]

The target fish in the grounds off Hong Kong was black marlin, while off the Philippines they fished for yellowfin tuna. In general, Siao Lioucciou fishers preferred the waters off the Philippines to the grounds off Hong Kong. This was partly because the waters off both the Philippines and Siao Lioucciou were rich zones with which they were thoroughly familiar. Equally important

[41]Chen Shengli, interview, 18 March 2002. The dolphin fish [*Coryphaena hippurus*] is not dolphin.

[42]Chen Shengli, interview, 18 and 21 March 2002; and Hong, interview.

[43]Chen Shengli, interview, 18 March and 24 June 2002.

was that the shallow waters of the South China Sea created larger waves and rough seas; hence, they did not like to fish there on a regular basis. Every year a large number of Siao Liouciou fishers left the South China Sea for the western coast of the Philippines to fish for yellowfin tuna between April and June. But due to the gradual thinning out of marine resources there, their search for new fishing grounds gradually moved eastward (see figure 7.5).[44]

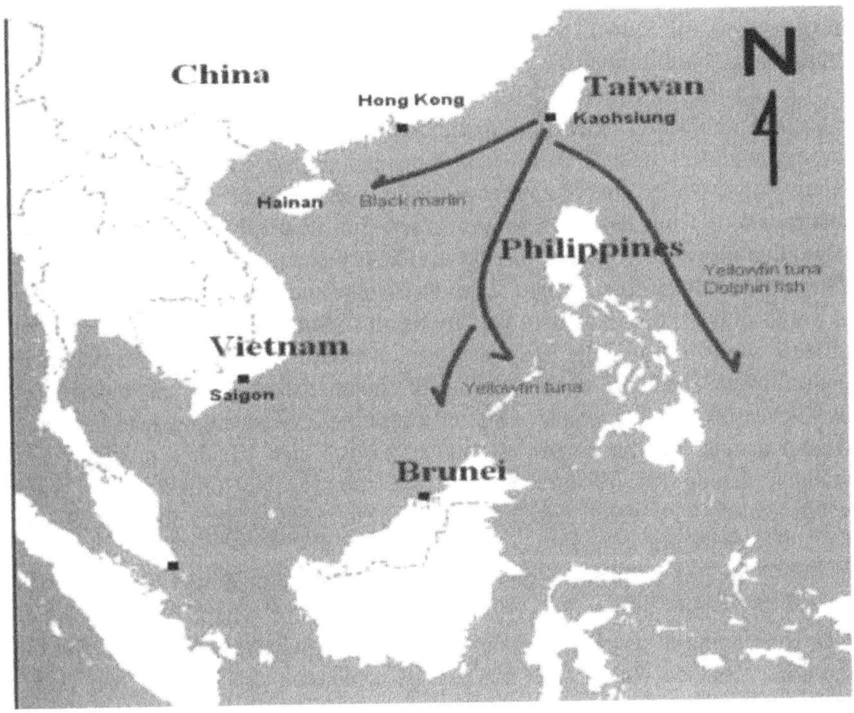

Figure 7.5: Fishing Grounds of Siao Liouciou Fishers after Moving to Kaohsiung

Source: See figure 7.1.

Crossing the Line: Detention and Confiscation

The detention of crews and vessels could occur under many different circumstances in foreign waters. First, fishers encountered both natural and manmade emergencies and took shelter overseas at foreign ports. For those Taiwanese fishers who worked off Southeast Asia, fishing in coastal waters or mooring in

---

[44]Chen Shengli, interviews, 18 and 21 March and 24 and 29 June 2002; and Hong, interview. Today a large number of Taiwanese fishers work the waters off the east coast of the Philippines.

foreign ports was considered a taboo which they would not break except in dire circumstances, such as engine malfunction or the approach of a typhoon, because the navies of some coastal nations could accuse them of illegally entering their territorial waters, poaching fish or smuggling.

For example, when *Baocing No. 1* was fishing in the open sea near Australia on 19 January 1975, its telecommunication system and engine simultaneously malfunctioned. The vessel drifted for two days and nearly capsized. Fortunately, with the help of *Ame-Rupe*, an Australian oil tanker, *Baocing* was eventually towed to Merauke on West Papua. The Indonesian authorities initially did not intend to seize *Baocing* and promised that it could leave after it was repaired. Local prosecutors, however, suddenly decided to confiscate the vessel, charging it with fishing in Indonesian territorial waters without permission. The court decided that the fishing master had illegally entered Indonesian waters and that he had poached marine stocks from the waters off Merauke. Hence, the vessel, gear and catch were confiscated, and the crew members were forced to leave Indonesia at their own expense.[45]

When Taiwanese fishers navigated through the territorial waters of various coastal countries to reach particular fishing grounds, that state's navy was liable to fabricate an unwarranted charge against them. To shorten the length of fishing trips, fishing masters sometimes risked a passage through the territorial waters of coastal states. The right of innocent passage was definitely a risky business. When fishers encountered the navy on patrol, the latter could easily charge them with poaching or smuggling. For example, in 1984 a set of Taiwanese pair-trawlers, *Menghong Nos. 21 and 22*, sailed through Indian territorial waters but did not fish. The Indian Coast Guard, however, believed that they had caught fish illegally because their nets were wet, even though the naval officers did not witness any fishing.[46] As a result, the vessels were detained.

A third reason for detention and confiscation was because fishers stole fish in the territorial waters of Asian coastal countries. Armed with large refrigerators and powerful engines, numerous Taiwanese fishing vessels illegally poached fish in the territorial waters of coastal Asia. *Jinhonghuei*, a Taiwanese offshore longliner, regularly defied the law and encroached upon the territorial waters of the Philippines. The vessel was caught by the Philippine Coast Guard and impounded in Iloilo on 28 April 1972.[47]

---

[45]Fu Kuncheng, *A Study of the Foreign Courts' Decisions regarding the Illegal Entry of Taiwan's Fishing Vessels* (2 vols., Taipei, 1990), I, 8.

[46]*Ibid.*, I, 18; and Defence Technical Information Centre, http://www.dtic.mil/doctrine/jel/doddict/data/i/02668.html (DOD), accessed on 15 March 2006.

[47]Fu, *Study*, II, 2-3.

Another reason for the detention and loss of vessels was that fishing companies failed to recognize or understand the proper procedures pertaining to international fishing agreements. The waters directly off the coast were the best fishing grounds for pair-trawlers. Before the 200-mile EEZ was introduced in the late 1970s, some Taiwanese fishing companies, in order to gain legal access to the inshore seas of coastal nations, attempted to reach international cooperation agreements. But without the full support of the Taiwanese government, they were easily intimidated. The following two examples help to illustrate this point.

In 1961, the China Fishing Company (CFC) entered into a fisheries cooperation agreement with Thailand. The two parties agreed upon a trial period of half a year. The CFC, however, unilaterally decided to terminate the arrangement before the contract expired because of poor port facilities and disputes over vessel registrations. This hasty decision incurred the wrath of the Thai authorities. *Yuping Hou* and *Yuguei Hou*, two Taiwanese single-trawlers, and all their crew members were detained in retaliation. Meanwhile, in 1975 the Yulong and Fumao Fishing Company negotiated a fisheries cooperation arrangement with P.T. Kaltim Hasil Laut, an Indonesian firm based in East Kalimantan. The Taiwanese fishing fleet, comprising *Feiteng No.1, No. 10, No.11, No.21 and No. 31*, as well as *Yulong No.2 and No. 10* and *Yadong No. 1 and No. 2*, docked at Santarina after obtaining permission from the East Kalimantan Provincial Government in September 1975. But tensions between Jakarta and the provinces meant that the deal with P.T. Kaltim Hasil Laut and the provincial government was not formally acknowledged by the Indonesian government. Consequently, P.T. Kaltim Hasil Laut was compelled to liquidate, and the large unsuspecting Taiwanese fishing fleet was suddenly detained by the Indonesian navy.[48]

To prevent detentions, fishing masters sometimes bribed foreign naval officers. This illegal practice, however, did not always work. Detention still occurred if fishers happened to come upon an honest official. On the other hand, so many Taiwanese fishers attempted to solve their overseas problems through bribery that it made the authorities even greedier. In countries like Indonesia, they became extremely eager to inspect Taiwanese fishing vessels. Some behaved like beggars, others like scoundrels. Su, a trawlerman who worked off eastern Indonesia, recalled that some naval officers simply extorted money vessel by vessel, and he felt they had no sense of shame. Such corrupt behaviour incurred the resentment and contempt of Taiwanese fishers.[49]

---

[48]"A Brief Introduction to State-Run Industries, 1961," in *The References of the Ministry of Economic Affairs* (Taipei, 1961), CCXXXII, 18-20; *Fisheries Tribune*, No. 8 (1962), 8, and No. 10 (1962), 3; and Fu, *Study*, I, 510.

[49]Fu, *Study*, I, 1-2; and Su Liansan, interview, 19 March 2002.

The fishing master and engineman were always separated from the rest of crew once they were detained because they were in charge of the vessel. The fishing master in particular had to endure an endless process of interrogation. Fishers were usually held in custody for a long period before the court tried their cases. The mental stress this created generally led the fishers to plead guilty; thus, the actual legal process rarely was longer than two months. In comparative terms, courts in the Philippines tended to respect the detainees' human rights the most, remained neutral and did not discriminate against foreigners. As a result, the fishers did not suffer the loss of their freedom and dignity to the same extent as their counterparts in Indonesia and elsewhere.[50]

In Indonesia, the attitude of law enforcement officials was at times both corrupt and harsh. Human rights were not respected. Nor was the legal process transparent. Court decisions could be changed repeatedly and without notice or explanation. The government at every level always discriminated against Taiwanese vessels. When the international fisheries cooperation arrangement between Taiwan and P.T. Kaltim Hasil Laut was rejected by the Indonesian government in 1975, the vessels that participated were confiscated by the Indonesian navy. Moreover, naval prosecutors accused the Taiwanese fishers of having entered Indonesian territorial waters illegally and of smuggling. In this case, the two charges were blatant fabrications. During the interrogation period the Taiwanese fishers were not allowed to hire lawyers or defend themselves in court. On the contrary, they were physically beaten until they pleaded guilty. The fishing fleet, which comprised nine modern, well-equipped trawlers, was then confiscated by the local authorities.[51]

Sometimes the Indonesian government, without prior notice, changed its mind about previous arrangements, which shocked the fishers and fishing companies concerned. *Baocing No. 1* is a good example of this erratic behaviour. Evidently, official permission from the Indonesian authorities to use the harbour at Merauke did not necessarily guarantee that *Baocing* was immune from being detained. Furthermore, Indonesian courts sometimes refused to hear cases that arose from fisheries disputes, insisting instead that fishing masters adopt a more conciliatory tone with the Indonesian government and the navy to reach an out-of-court settlement. But this "conciliation" process usually meant that the navy or local marine police could arbitrarily decide how to charge those fishing vessels that had legal or business concerns with the Indonesian government. The rights of fishers, as a consequence, were not protected.[52]

---

[50]Fu, *Study*, I, 1-2 and 9-10; and Jheng, interview, 13 March 2002.

[51]Fu, *Study*, I, 8-9.

[52]*Ibid.*, I, 8 and 488-489.

In more extreme situations, the confiscation of a vessel could force a company into bankruptcy. Therefore, most vessel owners were pleased to pay a fine if they could get their vessel back. In the early 1970s, the monetary penalty in the Philippines for Taiwanese vessel owners was comparatively small and did not create a financial burden. When *Jinhonghei* was detained by the Philippine navy in 1972, the owner simply paid 500 Philippine *pesos* to get his boat back. After a court decision, most Taiwanese fishing companies followed their lawyers or lobbyists' instructions and paid the fine, and few bothered either to read or contest the verdict.[53]

To repel Taiwanese fishing vessels (or to extort money from their owners), the fines increased dramatically and punishments became harsher. Sometimes vessel owners could not get their vessels back, even though they paid an enormous fine. Fishing masters and enginemen were considered the chief violators and were often put in jail. Other crew would be released on bail after their bosses sent the necessary money from Kaohsiung. The confiscation of a vessel was not only a tragedy for the vessel owner but also a major setback for the fishers, since the catch would be confiscated, thus depriving them of their profits. Often they needed to borrow money to support their families when they returned to Kaohsiung. Yet they could not necessarily expect to obtain financial support from a vessel owner who had just lost his vessel and might be facing bankruptcy. He would not have any sympathy under such circumstances for the struggling fishers.[54]

---

[53]*Ibid.*, I, 1 and 3.

[54]Jheng, interview, 13 March 2002.

Chapter 8

The Culture and Daily Life of the Kaohsiung Fishing Communities

As the fishing centre of southern Taiwan, Kaohsiung after the war attracted fishing migrants from Penghu, Siao Liouciou and other parts of the island. Hence, Kaohsiung offers a good opportunity to observe their migrant experiences and how socio-cultural factors like kinship and ethnic boundary maintenance affected their job prospects. Various fishing communities engaged in different kinds of fisheries-related activities. Therefore, I will examine both the spatial distribution of their economic activities and how the introduction of new fishing gear transformed the activities of fishing villages. As long-distance fishers, Taiwanese men were frequently away from home. In this context, it is interesting to explore how they managed to find wives and, after they were married, the kinds of marital problems that occurred when fishers were away for extended periods. Also of great importance is the fact that various types of fisheries offered different terms of employment. In the last section of this chapter I will illustrate the personal nature of the relationship between vessel owners and fishers and compare the employment and welfare practices in the different fishing sectors.

Migration, Job Seeking and Ethnic Boundaries

The Penghu Islands comprise several impoverished specks off the southwest coast of Taiwan (see figure 8.1). Traditionally, women and children remained onshore and worked the parched land which could only produce peanuts, sorghum, sweet potatoes, sponge gourds and aloes. The lack of agricultural productivity and sustainable natural resources drove Penghuan men to make a living on the sea. Hence, in terms of the gender division of labour, men worked at sea, while women worked on shore. The Penghuans called this cultural-ecological pattern "half mountain, half sea."[1]

[1]Chen Youyi's wife, interview, 6 July 2002. In most fishing communities, "the division of labour between the sexes seems in one respect quite sharp." See Paul Thompson, *et al.*, *Living the Fishing* (London, 1983), 173. Shore work is left to, or shared with, the women, but fishing at sea is monopolized by men.

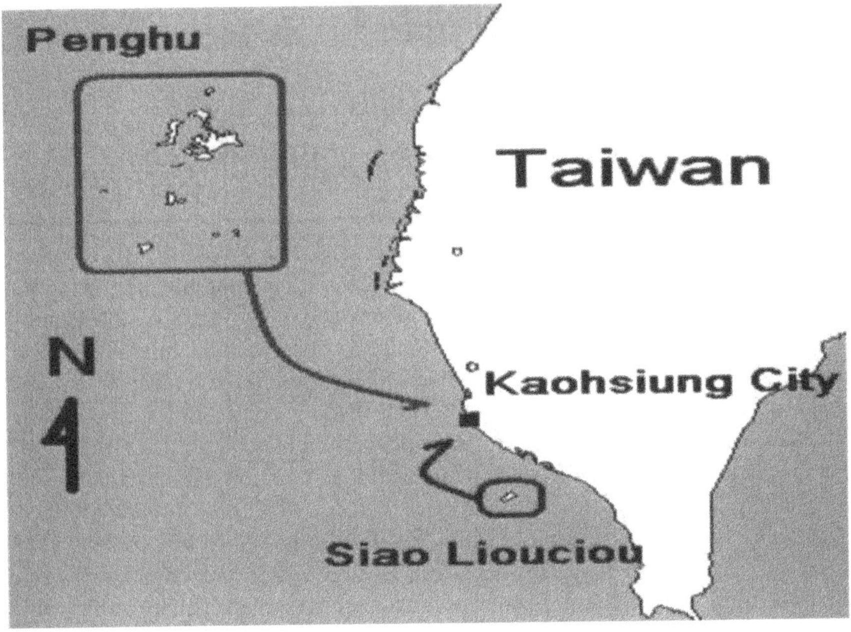

Figure 8.1: Penghuan and Siao Liouciou Fishing Migration

Source: Courtesy of the author.

The abundant marine resources of the Taiwan Strait, however, did not really contribute significantly to their economic improvement since the population was so small that the supply of fish always exceeded local demand. The more fish they harvested, the poorer their families became as the price declined. Hence, some Penghuans were forced to be seasonal fishers, oscillating back and forth between Kaohsiung and Penghu every year, while others migrated directly to Kaohsiung. The first wave of Penghuan migrants came mainly from Cimei. They began their migration during the colonial period and in general preferred to settle in Cihou rather than Gushan.[2]

The economy in Siao Liouciou was worse than in Penghu. The island is shaped like a shoe, and it is also extremely tiny: from northeast to southwest it is about 4.1 kms long, and from northwest to southeast just two kms wide. Its surface consists of limestone and coral sand which cannot sustain viable agricultural activities. Hence, the people were also forced to make a living at sea or to seek fisheries jobs in Kaohsiung.[3]

---

[2]Chen Youyi, interview, 6 July 2002.

[3]Chen Hsien-Ming, "The Development of Tuna Longline Fishery on Lioucniou Island," *Geographical Research*, No. 33 (2000), 201.

Even before Penghuan and Siao Lioucioou fishers migrated to Kaohsiung on a large scale in the postwar era, a number of fishers - mainly Penghuans - had worked as seasonal fishers in Kaohsiung in the prewar sailing-ship period. In the typhoon season they stopped fishing and returned home. In winter they came to Kaohsiung once again. There were two critically important dates in their yearly schedule; the first was the sixteenth day of the fourth month of the lunar year (May), a date traditionally viewed by fishers as the first day of the typhoon season. At this time of the year the weather became turbulent and the sea dangerously rough, as if, in the Taiwanese fishers' colloquial language, "the waves had opened their eyes." For safety, fishers remained onshore until the ninth day of the ninth month of the lunar year (October), a date that was regarded as the end of the typhoon season when fishers could safely return to the sea.[4]

The rapid growth of the Kaohsiung fishing industries at the end of World War II further encouraged Penghuan and Siao Liouciou fishers to settle there. The second wave of fishing migrants brought people from Siao Liouciou, Penghu and even Keelung.[5] As a result, seasonal migration gradually ceased. But living costs in Kaohsiung were much higher than in their home-towns, which made the destitute migrants feel even more impoverished. Most of the postwar migrant fishers dressed shabbily; some even went barefoot. They frequently had to seek refuge with their relatives after arriving in Kaohsiung and had to secure a position on a vessel as soon as possible. Before they could get a job, however, they needed to borrow money from their relatives to support their job search and their new way of life. Most of them, especially those from Penghu, were rural teenagers who did not readily understand the rituals and business practices of the modern urban world. If no one told them, they had no idea, for example, that they needed to offer gifts to fishing masters in exchange for a job. As a result, some youths merely wandered aimlessly looking for work.[6]

---

[4]Cai Minghuei's wife, interview, 5 July 2002; and Jheng Sanbian, interview, 20 June 2002.

[5]In fact, the migration of fishermen has occurred in many countries. What happened in nineteenth-century Britain is a good example. Devon trawlermen shifted northeastward to Humberside in several phases and are called "multi-stage migrants." Many Thames fishermen moved to Grimsby and settled there; they are called "single-stage migrants." Margaret Gerrish, "Following the Fish: Nineteenth Century Migration and Diffusion of Trawling," in David J. Starkey, *et al.* (eds.), *England's Sea Fisheries: The Commercial Sea Fisheries of England and Wales since 1300* (London, 2000), 112-117.

[6]*Ibid.*; Syu Yisin, interview, 4 July 2002; Chen Youyi, interview, 26 June 2002; and Jheng, interview, 20 June 2002.

Fishing masters preferred to employ relatives or those who offered presents or bribes. This employment practice was an "open secret" among those in the Kaohsiung fishing industry. When one of my elderly informants, Jheng Sanbian, tried as a young man to find onshore work in a fishing-related industry, he visited a number of establishments in Gushan and expressed his willingness to act as a courier in exchange for meals. Although he was not asking for a wage, no one was willing to hire him.[7]

But we can also examine another case with a rather different outcome. Syu Yisin, another Penghuan fisher, was far luckier in that he neither had to beg for a job or bribe any of the fishing masters in exchange for an onboard position because his uncle was a fishing master. In an employment-seeking situation like this, kinship connections mattered. Syu said that as an unemployed job seeker it made a big difference if you had kin relationships in the fishing communities. He also defended his uncle and this traditional kind of employment practice, saying that reserving a position for a relative or fellow villager was regarded as a "moral obligation." When Taiwan was in the grip of poverty in the 1950s and 1960s, onboard jobs provided relatively decent earnings. As a fishing master, Syu's uncle certainly felt he had to look after his "own people." Indeed, most of my interviewees said that getting a job onboard or a promotion required establishing good relationships with fishing masters or companies. Yet when they talked about their own cases, most stressed that their circumstances were unusual; they stood on their own feet, even though many had relatives who worked as fishing masters.[8]

This customary practice gradually disappeared in the 1970s when Taiwan's economic upturn offered young people from rural areas and offshore islands access to onshore job opportunities. In this economic climate, working at sea was no longer attractive, which led to major changes in the nature of interpersonal relationships between vessel owners, fishing masters and fishers.[9]

In the immediate postwar years, job opportunities in the fishing industry were so limited that the ethnic boundaries between different groups were fairly strict. When a fishing master recruited his crew, he invariably chose those who came from the same village. Therefore, it is safe to say that few Penghuan fishing masters hired Siao Liouciou fishers. Also, few Siao Liouciou fishing masters hired Penghuan fishers. One old Penghuan fishing

---

[7]*Ibid*. Offering presents or bribes to gain an onboard position also took place in Britain's trawling industry. In Hull, no fisher admitted that he offered a bribe to get his job, "but everybody knows that they are given." See Jeremy Tunstall, *The Fishermen* (London, 1962), 125-126. The situation was exactly the same in Kaohsiung.

[8]Syu, interview, 4 July 2002.

[9]*Ibid*.

master said that "[t]hey [the people from Siao Liouciou] had their own vessels, they could look after themselves." In fact, possessing their own boats was simply an excuse; the real reason was that they wanted to reserve job opportunities for people from their own ethnic group. As a result, neither Penghuan nor Siao Liouciou fishers would try to jobs with a fishing master from another ethnic group because they realized that the effort was useless. Only the radiomen could avoid the social problems generated by ethnic divisions and boundary maintenance because the shortage of skilled wireless operators meant that they did not have to defer to fishing masters to obtain a job. In fact, fishing companies had to recruit them by offering higher wages.[10]

Ethnic groups also clearly had different cultural ideas and practices in the selection of fishing methods. Siao Liouciou fishers did not want to be away from home for long periods, so they preferred to build small vessels and work on the offshore fishing grounds. On the other hand, Penghuan fishers were used to working on distant-water vessels and did not wish to work on the offshore fishing grounds or compete with local or Siao Liouciou fishers.[11] This preference for particular types of fishing methods further reduced the flow of fishers between the Penghuan and Siao Liouciou communities.

Still, it is important to stress that fishers from the same ethnic group did not necessarily cooperate. As mentioned in chapter 6, few Penghuans owned fishing companies in the prewar and early postwar years, except for the Chen brothers who had been quite successful in the Takao longline industry. Hence, they were expected to be kind and supportive to Penghuan migrants. Ironically, both were well-known in the Penghuan fishing community for their harsh and aloof behaviour.[12] Penghuan migrants came from various islands; most had strong ties to a particular place and often schemed for personal self interest and financial gain. Many cases of rivalry and influence-peddling can be found among the various Penghuan groups. They sometimes informed on their colleagues; old hands bullied novices; and they slandered each other out of jealousy and resentment. This kind of scenario was commonly viewed as part of the parochial nature of Penghuan fishing culture. To escape this, some Penghuans chose to distance themselves from their fellow fishers after they settled in Kaohsiung.[13]

---

[10]Chen Youyi, interview, 26 June 2002; and Jhong Ciourong, interview, 15 April 2002.

[11]Chen Youyi, interview, 26 June 2002; and Syu, interview, 16 April 2002.

[12]Jheng, interview, 20 June 2002. They also served as loan sharks when fishers needed to borrow money.

[13]*Ibid.*; Chen Youyi's wife, interview; and Chen Shuteng's wife, interview, 14 June 2002.

Onshore Economic Activities in the Kaohsiung Fishing Communities

When fishers were at sea, what kind of fisheries-related activities did their family members undertake in Kaohsiung? Different types of activities existed in various districts. Gushan had been urbanized since the colonial era and served not only as a distant-water fisheries centre but also as a hub of local commerce. Fishers' family members could therefore work in all sorts of modern industries. Thus, their economic activities were not necessarily related directly to fishing. The female relatives of fishers, living in flats, could work in processing factories, restaurants, stationery shops and grocery stores, or seek employment as stall-keepers or maids.[14] Cijin, however, was less developed and less prosperous. Most people living there were still engaged in the coastal fisheries and fisheries-related activities. The manufacture and maintenance of fishing nets or longlines, and aquaculture activities, such as collecting milk fish and preparing oysters, occupied most of their time.

Fishers could buy nets and longlines from gear shops, but these often were too expensive. To save money, some women in Cijin made fishing nets and longlines for neighbours. The nets they produced were small and could be used only in the coastal and offshore fisheries. Such nets and longlines rotted and tore easily, so women had to spend much time darning them. Some women were hired at low wages to do this arduous but important work, and the money they earned helped with the family's expenses.[15]

Fishing nets and longlines traditionally were made from bark thread, although some were woven from cotton yarn. The history of bark-thread fishing nets and longlines goes back much further than those made from cotton. Weaving and mending nets was a traditional skill handed down from generation to generation. The various types of bark were transported from Pingdong, a county at the southern tip of Taiwan. People in the fishing villages had to buy bark at local shops and manufacture nets or longlines on vacant lots in their village. The process of making the thread and weaving a net was very time-consuming. One informant, the wife of Chen Shuteng, explained that making bark thread required two steps. First, the bark had to be soaked in water. After it softened, it was torn into strips. Then the strips had to be exposed to the sun until they dried, after which they were twisted into strings. The production of bark threads was complicated because the strings often broke. Normally, women in the fishing communities undertook this task because it was not too demanding physically and required patience. In the 1950s, fishing nets and longlines made from bark thread gradually were replaced by factory-made cotton fishing nets and longlines which sold in

---

[14]Chen Shuteng's wife, interview.

[15]Cai Bian's female relative, interview, 3 July 2002.

fishing tackle shops at reasonable prices. Hence, women were eventually liberated from the burden of making bark thread and nets.[16]

Figure 8.2: Inhabitants of Taiwanese Fishing Community Exposing Fishing Nets to the Sun to Dry

Source: Courtesy of the Kaohsiung City Fisher Service Association.

Both bark thread and cotton yarn are plant fibres that easily rot in water. To lengthen the use of bark nets, people in fishing communities spent a lot of time on maintenance. Apart from darning fishing nets every day, they also had to clean and put a protective coating on the nets and longlines at least once a week. The protection of the fibre was an extremely laborious job that required cooperation from both genders. Women took charge of extracting the starch from red potatoes. They grated them on a wooden board embedded with nails, then mixed the paste with water before immersing the nets and longlines in the mixture. The fishing nets and longlines were then hung out on racks and exposed to the sun until the starch was totally dry (see figure 8.2). During this

[16]Chen Shuteng's wife, interview; and Cai Bian's female relative, interview. It is a universal truth that fishers' wives became involved in fisheries-related activities. I use the fishing communities on the Scottish east coast as an example. Fishers' wives in the 1970s still prepared the bait and mended the fishing gear for their husbands. See M. Estellie Smith, *Those Who Live from the Sea: A Study in Maritime Anthropology* (New York, 1977), 30-31.

period, people also had to collect pigs' blood from farms, dry it and grind it into a blood-red powder. The women then mixed this with water and immersed the nets and longlines to provide another layer of protection.[17] They then put the nets and longlines into a huge cauldron over a large steamer and heated them until they were totally boiled. Until the mid-1950s, men and women did this weekly so that the nets or longlines would not fray and become useless.[18]

Besides red potato juice, there were two other natural materials that could be used: egg whites and mineral pitch. Traditionally, people placed the longlines in a wood basin filled with egg whites and rubbed this thoroughly through the longlines, covering them with a protective layer of albumin. Then they wound the longlines onto wooden spools and exposed them to the sun to dry. The longlines were then also put into a cauldron and steamed. Maintaining the longlines with a coating of egg white was costly because eggs were expensive. Some longlines were also protected with mineral pitch, but this tar was much less viscous than the type used in macadam road construction. This technique was considered "advanced" in the 1940s and 1950s and applied only to the maintenance of distant-water longlines. Such longlines were stronger than those treated with potato extract and blood powder, and could effectively prevent both cotton yarn and bark string from corrosion by sea water. The catch was not affected if fishers used cotton longline dyed with mineral pitch, but it would decrease if they used bark-thread longlines treated with mineral pitch because when thread was coated with pitch it became as thick as electric wire and scared the fish away.[19]

The treating of longlines and fishing nets not only took place in offshore fishing communities like Cijin but also in Gushan. In the late 1950s, poor people in Gushan still treated longlines for vessel owners or fishing companies in exchange for fish. But since the 1960s nylon yarn has gradually become the norm in the industry (see figure 8.3).[20] Thus, people no longer

[17]Guo Shihfu, interview, 27 June 2002; and Thompson, *et al.*, *Living the Fishing*, 167-168. Women also played an important role in fish processing and preparing for the fishery elsewhere. In the UK, net-mending was a home task for fishers' wives and daughters.

[18]Guo Shihfu, interview; and Cai Bian's female relative, interview. The protection of fishing nets and longlines was a troublesome job. Therefore, some people in the fishing communities began to shorten the process by leaving out the pigs' blood because they felt that treating the nets with red potato starch was good enough.

[19]Guo, interview; Chen Shuteng's wife, interview; and Cai Bian's female relative, interview. In the 1940s and 1950s, longlines used in the waters off Southeast Asia were dyed with mineral pitch.

[20]Chen Shengli, interview, 21 May 2002; and Guo, interview.

needed to spend as much time on the maintenance of nets and longlines. The introduction of nylon to some extent changed the traditional outlook and employment opportunities for workers in the Kaohsiung fishing communities.

Figure 8.3: Woman and Child Preparing Nylon Fishing Nets and Gear at Home

Source: See figure 8.2.

Figure 8.4: Oyster Farm along the Beach of Cijin District

Source: See figure 8.2.

There were also numerous oyster farms along the beaches of Cijin District before the ports and harbours were renovated (see figure 8.4). These farms created many job opportunities. Before oysters could be cultivated in oyster beds, poles had to be fixed on the seabed so that seed oysters could be hung on the poles and cultivated. Some local people were hired to do this work using bamboo rafts from which they handed the special poles to divers. The majority of the divers were new migrants from Penghu because local oyster farmers believed that Penghuans could hold their breath for a long period. Divers took five or six bamboo poles with them on each descent and would not emerge until all had been securely fixed on the seabed. Oysters would then have to be harvested before the onset of the typhoon season.[21]

Figure 8.5: Women in the Cijin District Shucking Oysters

Source: See figure 8.2.

Women in the fishing communities were also kept busy shucking oysters (see figure 8.5), which were an important ingredient in *zongzih*

[21]Cai Minghuei's wife, interview. In autumn and winter, oyster farmers fixed poles on the bottom of the water and harvested oysters several months later. See *The Fishing Industry of Taiwan* (Nantou, 1953), 11.

(traditional pyramid-shaped rice dumplings). To preserve the oysters, the women had to spread them out in the sunlight and rotate them regularly until they were completely dry. Peddlers would then come to collect the dried oysters for the annual Dragon Boat Festival, held on the fifth day of the fifth month of every lunar year, when there was a huge demand for dumplings.[22]

Women in the fishing communities were also kept busy collecting baby milkfish two weeks prior to the annual Tomb Sweeping Day (around 5 or 6 April). They waded into the water and used small nets to collect the tiny fish. How many they could collect depended on the weather. Usually they could anticipate a better catch in the season of the south wind. Milkfish collectors came to each village to collect the catch. A woman might earn between NT $100 and NT $200 in a single day, which was considered an extraordinary sum.[23] Even some of the coastal and offshore fishers in Cijin halted their regular activities and concentrated solely on milkfish collection during the typhoon season. At this time, milkfish collection was safer and more lucrative than fishing. Women in the fishing communities worked on milkfish collection until the advent of White Dew (around 8 September), after which both the quantity and quality of the tiny fish gradually decreased.[24]

Postwar fish-processing factories in Gushan also provided a lot of job opportunities for women. The female relatives of fishers were hired to shuck shrimps caught by offshore trawlers. The shrimps they shucked were then processed and exported to Japan. Most of these workers were casual labourers recruited by a headwoman. This casual work was extremely competitive, so those who wanted to keep their jobs had to be expert and efficient or have a very good working relationship with the headwoman.[25] One of my elderly female interviewees, Mrs. Syu, recalled how extremely competitive the job was: "When I was in my early teens, very often I had to take a stool and then sit and wait in the queue before dawn."[26]

---

[22]Cai Minghuei's wife, interview.

[23]At that time a decalitre of rice cost a NT $12. See *Fishing Industry of Taiwan*, 11.

[24]Cai Bian's wife, interview, 3 July 2002; and Cai Minghuei's wife, interview.

[25]Chen Youyi's wife, interview; Chen Shuteng's wife, interview; and Syu Yisin's wife, interview 6 July 2002.

[26]Syu Yisin's wife, interview. In the UK, women also made a significant contribution to fish processing. They worked as gutters and whitefish filleters in fish processing factories; some worked as kipperers in smokehouses. Like fishers' wives or daughters in Taiwanese fishing communities, they were hired seasonally, and their wages were very low. See Thompson, *et al.*, *Living the Fishing*, 168.

Besides working in the fish-processing industries, cleaning vessel hulls was another job option for people in fishing communities. After a long trip at sea, moss and barnacles grew on vessel hulls. To maintain a fine exterior, vessel owners had the hulls cleaned regularly when the boats returned to Kaohsiung. This was not a job which required much physical strength, and it usually was delegated to members of the fishers' family, such as young sons or even daughters. This unique scene from the Kaohsiung Fishing Port gradually disappeared in the 1970s when vessel owners began to entrust this important work to hull-cleaning professionals.[27]

Marriage and Family Problems

In most cases, marriages of young fishers were arranged by their parents who, with the help of matchmakers, looked for a prospective woman of the right age and then arranged a meeting when the son returned from fishing. Young fishers rarely had much choice except to marry a woman from the same social background because few young women outside the fishing communities were willing to marry them because of their poverty and low social status. Thus, the fishers' brides came mostly from the Kaohsiung fishing communities or their hometowns on Penghu or Siao Liouciou. Normally, the two parties would get married soon if their arranged meeting was a success. Some couples held their weddings in their hometown, some in Kaohsiung. It depended on the personal circumstances and financial means of the two parties.[28]

Still, sometimes a poor but talented fisher could marry a wealthy young woman. If the young man was considered trustworthy, industrious and promising, he might be able to marry a fishing master's daughter. Occasionally, some fishers even married vessel owner's daughters and then experienced a meteoric rise in the industry. As Syu Yicin explained, "[i]t was a natural thing that fishing masters introduced their daughters to you, if you were a competent and promising man."[29]

---

[27]Syu Yisin's wife, interview.

[28]Chen Shuteng's wife, interview. In Asia, marrying a person from the same social background does not happen only in societies like conservative Taiwanese fishing communities. It also happens in the Western world. The Italian fishing community in Fremantle, Western Australia, is a good example. Italian migrants still preferred to marry a person in Fremantle's Italian community or to bring a bride from Italy. See Charles Gamba, *A Report on the Italian Fishermen of Fremantle* (Perth, 1952), 56-57. The same thing was also found in Hull. Most of the young fishers met prospective partners "through the inter-connected webs of relatives, neighbours and friends." See Tunstall, *Fishermen*, 141-142.

[29]Syu, interview, 4 July 2002.

But this social practice raises an interesting question: did young women in fishing communities necessarily want to marry a fisher? Not really, at least from the standpoint of social mobility. Their fathers were fishers, and so were their brothers. To these prospective brides, marrying a fisher was no more than an acceptable option. In their hometowns, most had no choice but to marry a local fisher. After moving to Kaohsiung, however, young women had more potential suitors from which to choose and more opportunities to meet young men from outside the fishing environment. Often they would not marry a fisher if they found a more suitably employed man.[30]

"If you marry a fisherman, you only have a half husband," stated Cai Bian's aunt.[31] Husbands in long-distance fishing families were frequently away at sea for at least half of their married lives. Thus, wives had to look after the household and raise the children largely on their own. They had to be self-reliant, strong and independent. Hence, in a Taiwanese fishing family, the real person in charge was the wife rather than the husband, although he was the main breadwinner. If women encountered any personal or financial difficulties while their husbands were at sea, they often asked for help from their parents or siblings rather than their in-laws. Furthermore, women in the Kaohsiung fishing communities were usually inclined to support and help each other because all their husbands were away at sea. Compared to married women in other onshore communities, those in fishing settlements were not as close to their husbands' families, even though the patriarchal pattern of Chinese families and kinship was strongly reflected in traditional Taiwanese culture.[32]

In general, fishers' wives can be categorized into two different types. Some were industrious and thrifty in managing their households, and these women raised their children with an exemplary sense of duty. Others, however, simply could not endure the hardships and loneliness, forsook their families and either eloped with lovers or had extra-marital affairs. It is clear that the majority of fishers' wives conformed to the monogamous norm, but the number of "bad women" ought not to be underestimated. In some extreme cases, desperate housewives sold their homes and property, left the children with their grandparents and fled with their new partners. Their unsuspecting husbands, who were working at sea, had no idea what had happened until they belatedly arrived at Kaohsiung.

---

[30]Chen Shuteng's wife, interview.

[31]Cai Bian's female relative, interview.

[32]Cai Bian's female relative, interview; and Chen Youyi's wife, interview. Married women in the fishing communities were not very close to their husbands' families. A similar case can be found in Hull. When fishers were working at sea, their wives always sought comfort and assistance from their mothers rather than from their husband's family. See Tunstall, *Fishermen*, 161.

None of the fishing communities in Kaohsiung were immune from such social problems and family tragedies. It is difficult to say which fishing community was more likely to suffer this kind of personal misfortune. When discussing this sensitive issue, fishers always stressed that women from other communities were more apt to have extra-marital affairs than those in their own. A Penghu engineman stated: "Yes, I admit that some wives in our community had affairs with other men. But the chance was much lower compared with other groups."[33] Yet from an impartial perspective, this statement does ring true. Extra-marital affairs rarely occurred before families migrated to Kaohsiung. Both the Penghu Islands and Siao Liouciou were tiny places where people knew each another well. It was next to impossible to have an illicit affair in such a closed and conservative society.[34] But this was not necessarily the case after they migrated to Kaohsiung, the second largest city on Taiwan, with so many strangers and new opportunities.

In addition to having affairs, other temptations, such as gambling, also led women into disrepute. In their leisure time, some Penghuans liked to play games of chance, such as "four colour cards" (*sihsepai*) or *mahjong* with small wagers. Most of their playing partners were close neighbours or fellow villagers. To them, playing cards or *mahjong* was a form of recreation rather than an addiction because it seldom led to serious financial disputes or family problems. But this was no longer the case after the fishers migrated to Kaohsiung. Gambling in this urban environment became a serious problem for some Penghuan families for several reasons. First, numerous Penghuan migrants worked in the distant-water fisheries, so husbands were not home as often as when they lived in Penghu. Some Penghuan wives began to fill the void in their lives by gambling. The monthly family payments from the fishing companies became their main source of money for gambling. The greater the individual stipend they received, the more risks they could take. In these cases, playing cards or *mahjong* was now more than simply a recreational activity; it became a serious, albeit risky, habit that caused major family problems. Not surprisingly, this kind of household tragedy happened more often to fishers who worked on distant-water longliners than to those working on pair-trawlers. The trawlers returned home every two or three months, so the fishers had a better sense of the well-being of their families.[35] One retired Penghuan

---

[33]Syu, interview, 16 April 2002.

[34]Syu, interview, 4 July 2002.

[35]Cai Minghuei's wife, interview; Chen Shuteng's wife, interview; Cai Bian's female relative, interview; Chen Youyi's wife, interview; and Jhong, interview. Regarding the family payment, there was a similar system in Britain. When fishers were at sea they arranged for a regular weekly payment to be sent to their families. There-

fishing master said that among people in his community, as many as half were addicted to gambling. In the most extreme cases, apartments and houses had to be sold to repay gambling debts. Some wives even worked as dance hostesses at bars to earn money to gamble or pay off debts. Most of their husbands were not aware of what had happened to their wives and families until they returned to Kaohsiung from a fishing voyage, by which time it was often too late.[36]

Relations among Vessel Owners, Fishers and Families: Wages and Welfare

In the old days, some fishing companies signed contracts with fishers, and some did not. It depended on the vessel owners' whim. Their contract period initially was only for one year, but with the continuous expansion of fishing, contracts were signed for two years, and some were even for three. At the end of a voyage the catch was sold, the operating expenses paid and the net profits shared by the "people" and the "boat" (the fishers and the fishing companies). Fishers could receive up to forty-five percent of net profits, while the fishing company retained the balance. But this ratio was only a general principle; in practice, profit sharing varied from company to company.[37]

I will use the Siao Lioucíou profit-sharing system as an example. Here the vessel owner took half of the net profit and the crew shared the rest. Suppose there was a Siao Lioucíou vessel with a crew of eight; the catch of a fishing trip was sold for NT $ one million; and the running expenses were NT $200,000. In this case, the net profit would be NT $800,000. Hence, the vessel owners earned NT $400,000, while the eight fishers, including the master, shared the remaining NT $400,000. A fishing master and an engineman on a Siao Lioucíou vessel received double the share of ordinary fishers, which means that the NT $400,000 had to be divided into ten units, as the fishing master and engineman earned NT $80,000. Besides their normal income from this profit-sharing system, the fishing master sometimes could negotiate for a bonus, and in most cases the vessel owner agreed.[38]

In general, the larger the vessel, the bigger the share of the net profit that went to the fishing company. For distant-water longliners or pair-trawlers that worked on foreign fishing grounds, the fishing companies might take sixty percent of the net profit after a voyage. Also, the larger the vessel, the greater the bonus the fishing master and other leading crew could earn. While a Siao

---

fore, fishers' wives were slightly wealthier than their neighbours. See Tunstall, *Fishermen*, 160-161.

[36]Chen Youyi's wife, interview; and Syu Yisin's wife, interview.

[37]Syu, interview, 4 July 2002; and Jheng, interview, 13 March 2002.

[38]Hong Fucai, interview, 21 May 2002.

Lioucliou fishing master's bonus was double or three times that of ordinary fishers, the salary bonus of a Penghuan fishing captain who worked in the Pacific or Atlantic could be ten or even twenty times larger than the shares of ordinary crew.[39]

Before 1965, the profit-sharing system in the Taiwanese fishing industry was not transparent. The fishing master was the only person who could negotiate with the vessel owner over the exact amount of bonus that each fisher should get. This was partly because a fishing master, unlike other crew members, knew the exact amount of the catch and its possible market value; hence, his boss could not cheat him. It was also partly because some masters were also stakeholders in their companies, and the vessel owners had to invite them to internal company meetings. But there was little participation in this profit-sharing process by the fishers. As Chen Youyi explained, "[y]ou are not allowed to argue about the bonus with your bosses! If you do so, you will be asked to take a rest onshore."[40] Taking a rest onshore actually meant dismissal, because both vessel owners and fishing masters became concerned if fishers learned too much about the financial conditions of the fishing company or how the profit-sharing system operated behind the scenes.

Did the fishing companies actually look after the welfare of the fishers' family members? The answer is a qualified yes. Fishers spent most of their time at sea; fishing companies or vessel owners therefore had to look after their family members for the duration of the voyage. But this did not mean that vessel owners had a direct hand in the personal affairs of the fishers' families. Vessel owners and fishing companies simply offered regular financial support in the form of monthly payments and loans if necessary.

On a distant-water longliner, a fishing trip might last for more than one year. To assist fishers to support their families, each month longline companies provided a family allowance. If they lived in Kaohsiung, family members had to go to the company office to collect their stipend; if the families were located elsewhere, such as on Penghu, the companies were obligated to send the money by mail. Usually the family stipend was just enough for the fishers' wives and children to live a simple life. In addition, if a fisher or his family experienced a financial problem, they could borrow money from the company. The debt had to be repaid after the net profits had been shared upon the completion of a fishing trip or before the debtor left his position for another company. It is important to note, however, that borrowing from the company was just a customary practice rather than part of a normal system. If the employer was harsh, or if a fisher or his family did not have a good enough reason, he could simply ignore the request. Since trawl fishers

---

[39]Chen Youyi, interview, 26 June 2002.

[40]*Ibid.*

returned to Kaohsiung every two or three months, the trawling companies did not provide their families with an allowance. But to help fishers provide for their families, trawler owners had to advance money to fishers before they embarked on their voyage. Unlike the longline industry, this cash advance was part of a system rather than a practice. The advanced payment or assisted loan became one of the most important reasons for a fisher to work in the trawling industry. As Jhong Ciurong stated, "[i]t is easier to borrow money from a trawl fishing company." How much money fishers could borrow depended on how much money they could possibly earn on each trip. Normally, the amount of an advance would not exceed the fisher's potential earnings. For example, a radioman was usually able to borrow up to seventy-five percent of his estimated salary for a particular trip before going to sea. Some fishers attempted to take advantage of the practice and borrow as much as they could. Both trawling and longline companies would hold back such advances or deduct money from their bonus.[41]

In rare cases, fishers borrowed from their companies and then absconded, never returning to the community. Most of these men were raw recruits who had just signed contracts but regretted having done so even before the voyage began. They simply chose to disappear, stealing the money they had just borrowed. This was a criminal act, but taking the matter to court was complicated, and some vessel owners simply chose to drop the matter.[42]

Most important, a fisher could not always make enough money from a fishing trip to repay the money he borrowed. The debt was still there at the end of the voyage, but he then had to borrow from the company again in order to live before he started his next trip. As a result, his debt could accumulate, and some men just could not overcome their financial difficulties. To extricate themselves from such an invidious cycle, some indebted fishers attempted to work for other companies without repaying their debts. According to the practice in Kaohsiung fishing communities, however, an indebted fisher was not allowed to work for another vessel owner before repaying all his debts. Still, job-changing happened quite often; some people arbitrarily terminated their contract for a better position with another company. This practice was acceptable as long as the fishers could repay the money they had borrowed from their ex-employer. In some cases, the new company or vessel owner offered to help and incurred the cost of liquidating the debt.[43]

Financial disputes between fishers and vessel owners happened all the time. Shouting on the phone or in the company office over financial matters

---

[41]Chen Youyi's wife, interview; Jheng, interview, 13 March 2002; and Jhong, interview.

[42]Jhong, interview.

[43]Chen Youyi, interview, 26 June 2002; and Syu, interview, 4 July 2002.

was considered normal. In the early 1960s, a tycoon in the Kaohsiung fishing industry was stabbed to death by one of his fishers after a violent quarrel in the company office. In another case, the mistress of the Head of the Kaohsiung Fishermen's Association was killed by a fisher because of an argument over financial matters.[44]

In fact, from the perspective of capital, offering financial help to the fishers' families had another important function – namely, social security. Vessel owners or fishing companies could readily abrogate their financial and moral responsibility if an accident happened at sea. In most cases the fishers' family members would do everything possible to demand compensation from vessel owners or fishing companies when a shipwreck or detainment occurred abroad. They went to the company offices or vessel owners' residences and cried, shouted and made an embarrassing public row until they received a satisfactory settlement.[45] Traditionally, most vessel owners did everything they could to block these requests. They could claim that the fisher had received decent earnings through the share system. In addition, the company might have provided extra financial support to the fishers' families each month or on the occasion of a new fishing trip. The company would appear to have done everything it needed to do. The government, courts and even public opinion would generally side with the company if a compensation case went to court.

On the other hand, all the offshore longliners in Cijin were run by fishing families rather than by companies. They raised money among relatives, neighbours and friends to build small vessels to fish in the waters of the South China Sea or the fishing grounds off the Philippines. A Cijin vessel owner and stakeholders might work on the same vessel with fishers; therefore, the boundary between social obligation and duty was blurred. The owners and stakeholders did not provide a family allowance to fishers because the vessels returned home often. Moreover, they did not advance money to fishers because the vessel owners were themselves not much better off. In this case, the financial situation of a vessel was far more transparent since the fishers worked together with their boss. After returning to Kaohsiung, they sold their catch in the fish market together to further facilitate transparency.

Ironically, in Cijin, when an accident happened at sea, fishers' family members could not claim compensation from the vessel owner because the owner was invariably onboard and perished along with the crew. Social relationships in such circumstances were always far more harmonious. All the fishers knew each other well because for generations they had been relatives,

---

[44]Chen Youyi, interview, 26 June 2002; and Jheng, interview, 13 March 2002.

[45]Gu Tingfang, interview, 23 June 2002.

neighbours or friends. Hence, they trusted each other at sea, and their families customarily looked after each other on shore.

Nevertheless, vessel owners enjoyed all sorts of financial advantages, which encouraged some masters to set up their own fishing companies. Some succeeded, while others failed. The situation varied within each ethnic group. Comparatively, Penghuan fishing masters had the least chance of success.[46] Penghuans from a Buddhist background believed that certain Penghuan vessel owners had taken too many fish while working as fishing masters. The bankruptcy of their fishing companies and the financial difficulties they suffered were seen as just retribution for what they had previously done. Nonetheless, there is another more practical explanation concerning the failure of Penghuan entrepreneurship. Penghuan fishing masters spent half their lives at sea. Onboard a vessel, the fishing master was often the most experienced fisher, but he rarely could become a talented manager onshore. These men knew nothing of the management side of a fishing company and were unfamiliar with the business practices in Kaohsiung.[47] Their limited business network could not compare with those of Kaohsiung's local people. Local vessel owners tended to diversify their capital in many ways, and the fishing industry was just one of their investments; by spreading their risks, they increased the potential survival of their overall enterprise. But for most Penghuan vessel owners, the fishing vessel was their only major financial asset. Each might only own one or two distant-water longliners that were the fruit of their painstaking labour over a lifetime at sea. To some degree, their fishing companies were based on an extremely fragile economic foundation. If something major went wrong, such as a vessel confiscation or a shipwreck, they would go bankrupt.

Most Siao Liouciou vessel owners had also previously worked as fishing masters. But their situation, compared with Penghuan vessel owners, was much better. First, most owned and operated a fleet of small longliners rather than a single large distant-water vessel. This business strategy effectively reduced the risk of bankruptcy through confiscation or shipwreck. Second, operating small longliners in the nearby waters off Southeast Asia was technically and financially easier than running a huge distant-water vessel that worked on remote grounds in the Indian or Atlantic oceans. These reasons help explain why it was easier for a Siao Liouciou fishing master to become a successful vessel owner than it was for a Penghuan master.

Furthermore, people in Cijin maintained the strongest sense of a closed corporate fishing community. Cijin was always a less-developed area,

---

[46]I have to stress that today many fishing companies are run by Penghuans.

[47]Liou Syuan, interview, 4 July 2002; Syu, interview, 22 June 2002; and Sia Fujhong, interview, 19 June 2002.

but people in the villages knew each other well. Men worked together at sea, and their wives worked together and looked after each other in the villages. The mainstream fishing activities in Cijin were aquaculture and coastal and offshore fisheries. Women, children and the elderly participated in the manufacture and maintenance of fishing gear and provided all sorts of "logistical services." By comparison, the sense of community among fishers was weaker in Gushan. First, the distant-water longline and trawl fisheries of Gushan required more professional behind-the-scenes service, which prevented fishers' family members from getting directly involved. Second, Gushan had become extremely urbanized (see figure 8.6), and most fishers' families lived in apartments or in narrow alleys of urban districts and regularly mingled with people engaged in onshore business activity. Third, a variety of onshore industries in Gushan and its adjacent areas offered family members job opportunities. Working in the ancillary industries no longer seemed attractive in such a rapidly developing urban environment (see figures 8.7 and 8.8).

Figure 8.6: Early Photograph of Gushan District

Source: See figure 8.2.

Figure 8.7: Gushan Fishing Port in the Twenty-first Century (01)

Source: Courtesy of the author.

Figure 8.8: Gushan Fishing Port in the Twenty-first Century (02)

Source: See figure 8.7.

Cianjhen Fishing Port was completed in 1967, which encouraged numerous distant-water fishing companies to shift their offices and fleets from Gushan to Cianjhen. The government's key fisheries-related institutes and ancillary industries, such as ice-manufacturing, cold storage and fish-processing plants were now clustered near the port and formed a self-contained environment for the future development of Taiwan's distant-water fishing industry (see figure 8.9). Nevertheless, it would be incorrect to regard Cianjhen as a *bona fide* fishing community. First, fishers and their family members still resided in Gushan, Cijin or other parts of Kaohsiung; they rode their motorbikes to Cianjhen to board their vessels and did not need to move to the area. Second, when fishers bought second properties, a large number still chose downtown districts in Kaohsiung rather than Cianjhen. Their reason was often personal and had little or nothing to do with the fishing industry. Moreover, family members sometimes had jobs in the neighbourhood and were reluctant to relocate. Further, they did not participate in fisheries-related activities in Cianjhen as they had done in Cijin or Gushan. In other words, large-scale migration of fishers did not occur at Cianjhen as it did at Gushan.[48]

---

[48]*The Fishing Ports of Kaohsiung City* (Taipei, 2003), 9. The same thing happened in Scottish fishing communities. Fishers moved to modern houses, "reached their ships by car" and did not have to live in the port area. As a result, the fishing community was not as isolated as before. See Smith, *Those Who Live from the Sea*, 30.

Figure 8.9: Kaohsiung in the Twenty-first Century with the Cianjhen River in the Foreground

Source: See figure 8.7.

The overall sense of a fishing community was also further weakened by the rapid growth of Taiwan's economy. The renovation of ports and harbour facilities in Cijin, and the environmental pollution generated by Kaohsiung's heavy industries, to a large extent prevented fishers' family members from participating in traditional aquaculture activities or the coastal fisheries. Besides, the introduction of nylon fishing gear liberated people in the fishing villages from the manufacture and maintenance of nets and longlines. The surplus labour force in these fishing communities was soon absorbed by new employment in the nearby industrial parks. Young men could now earn a decent wage onshore without enduring the risks and hardships of life at sea. They preferred to become hairdressers, taxi drivers, construction workers or government functionaries rather than work at sea and sacrifice the quality of their family lives. The majority of fishers' sons increasingly refused to follow in their father's footsteps, especially after they moved to the downtown districts of Kaohsiung. The traditional practices that existed in the fishing villages, and the customary economic activities in which people engaged previously, gradually disappeared.[49]

The sense of a fishing community was also further eroded by the changing marriage patterns among fishers' children. In the old days, a fisher could easily marry a young woman from his fishing community. Through a pattern of intermarriage and kinship, the sense of community was maintained

[49]Few of the sons of my interviewees now work as fishers.

and strengthened. This gradually changed with the rapid development of Taiwan's modern economy. More young women in fishing villages wanted to leave for the city and to marry a man who worked onshore. For them, marrying a fisher was no longer considered desirable.[50] The future well-being of the fishing communities required the participation of both sexes in marriage and the workplace. But few in Kaohsiung, even fishers' daughters, want to marry fishers today. Young fishers have been forced either to abandon their fishing career or to marry a woman from Southeast Asia or China. Their children are not likely to inherit their father's fishing career. By the 1980s, the fishing communities in Kaohsiung were dying.

The lack of a viable local labour force for the fishing industry also reshaped the relationship between vessel owners and fishers. Due to numerous job opportunities in onshore industries, young men did not want to work at sea as much as in the past. In 1975 alone, sixty-five percent of Taiwanese distant-water fishers left their positions for onshore employment.[51] The recruitment of new fishers became onerous for fishing masters, who increasingly devolved the task to their companies. The fishing companies and vessel owners had no choice but to undertake this job. To continue to attract young people to work on the vessels, fishing companies and vessel owners had to offer better wages and family allowances to fishers and their families. Suddenly the situation became very attractive for ordinary fishers; the previous harsh attitude of many vessel owners and fishing masters and the culture of bribery in the fishing communities rapidly became history in the face of a booming Taiwanese economy and a rapidly dwindling fishing labour force.

---

[50]*News.yam.com*, http:yam.udn.com/yamnews/daily/2912495.shtml, accessed 11 September 2005.

[51]Wang Anyang and Liou Sijiang, *The Problems of Taiwan's Distant Water Fisheries and Research on Fisheries Policies* (Taipei, 1979), 32.

Chapter 9

The 1970s Crisis in the Taiwanese Fishing Industry

The late 1970s was a crucial period for Taiwan's distant-water fishing industry as the traumatic effects of a variety of serious problems were manifested simultaneously. Some of the problems were not preventable. The acute labour shortage in the distant-water industry, the energy crisis of 1973 and Taiwan's diplomatic isolation which led to the failure of cooperation arrangements in the international fisheries were three key examples. Some problems, however, were caused by the short-sightedness of Taiwanese business culture. The rapid depletion of regional marine resources and the relatively small scale of the fishing companies and their financial fragility were two important adverse consequences that the Taiwanese vessel owners brought upon themselves. While these problems developed over time, they came to a head in the late 1970s when the 200-mile Exclusive Economic Zone (EEZ) was claimed by increasing numbers of coastal nations.

Labour Shortage

In the 1940s and 1950s young fishers were forced by economic circumstances to bear the ill-treatment and loneliness of the sea because for most of them working on a distant-water vessel was the only way to make a living. In the 1970s, however, numerous job opportunities created by onshore economic activities in the metropolitan areas, particularly in Taipei and Kaohsiung, encouraged young people to leave the fishery. In 1973, a novice fisher working on a distant-water vessel earned an average of NT $3000-7000 (US $79-184) per month, which was not much better than the salary that a young man could make in an onshore factory. Hence, in that year thirty-eight percent of the distant-water fishers quit; in 1974 fifty-two percent left; and in 1975 sixty-five percent abandoned their jobs at sea. The employment situation continually deteriorated over time.[1]

Before the 1970s, getting a position on a vessel was quite competitive. In addition to having a good relationship with the fishing master, fishers had to

[1]*Fisheries Tribune*, No. 103 (1973), 3; and Wang Anyang and Liou Sijiang, *The Problems of Taiwan's Distant Water Fisheries and Research on Fisheries Policies* (Taipei, 1979), 32.

be industrious, skilful and obedient. By the 1970s, however, this was no longer the case because working at sea was not as attractive a vocation as before. Those who had even a slight knowledge of the fishery could now work in key staff positions. As for ordinary fishers, the companies could no longer expect them to be knowledgeable about fishing techniques. Some fishers were fugitives or ex-convicts. The lack of seafaring ability and navigational experience forced new fishers to rely heavily on modern technological devices. Furthermore, the decline in the overall calibre of the fishers indirectly increased the risk of marine accidents.[2]

Fishers with criminal backgrounds were frequently inefficient and disobedient when working at sea. Sometimes they engaged in gang fights or drunken brawls in foreign ports, and they invariably damaged Taiwan's reputation. As well, the level of education of fishing masters in the 1970s and 1980s was not much better than in the 1940s or 1950s. Most of the masters graduated only from primary or junior high school. Hence, they often had little idea of how to cope with emergencies in foreign waters or ports. According to a 1971 report in the *Fisheries Tribune*, forty-five Taiwanese offshore and distant-water fishing vessels had marine accidents, killing 113 fishers. The lack of education paints a disturbing picture. Only three of the deceased had graduated from senior high school; another three from junior high school; and the rest did not go beyond primary school. Some were illiterate.[3]

Graduates of fisheries vocational schools increasingly did not want to work at sea; they considered themselves members of the intelligentsia and refused to mingle with ordinary fishers. Some schools asked students to complete their apprenticeship at sea as a graduation requirement. In extreme cases, students who did not want to fulfil this condition begged vessel owners to provide a false completion certificate. The fishing companies were also reluctant to give these bookish young men the opportunity to fulfil their apprenticeship on their vessels because such "scholars" always questioned orders and had great difficulty coping with the hardships of being at sea.[4]

---

[2]Jhong Ciourong, interview, 15 April 2002; Huang Sianchih, "Economic Development and the Structural Adjustment of the Production of Taiwan's Fisheries" (unpublished paper presented at the Joint Annual Meeting of ROC Agricultural Groups, Taiwan, 1984), 8; *Fisheries Tribune*, No. 92 (1971), 12-13 and No. 97 (1971), 13; and Chen Youyi, interview, 26 June 2002.

[3]*Chinese Fisheries Weekly*, 26 July and 9 August 1976; Chen Zaifa, "Problems in the Development of Ocean Fisheries and Their Solutions" (unpublished Kaohsiung Fishing Boat Commercial Guild [KFBCG] internal document, n.d.), 17; and *Fisheries Tribune*, No. 97 (1971), 12.

[4]Gu Tingfang, interview, 1 July 2002; *Fisheries Tribune*, No. 92 (1971), 20, and No. 97 (1971), 12.

The lack of a decent labour force created two additional problems for trawler companies. Trawlers remained in port for shorter periods than other vessels; in the 1940s, 1950s and early 1960s, pair-trawlers stayed at Kaohsiung for only three to five days to take on sufficient supplies of fuel, food and other necessities before returning to sea. In those years, an adequate labour force was not a problem. In the 1970s, however, trawlers had to remain in harbour for longer periods in order to recruit fishers for the next trip. In addition, the fishers' economic stature and bargaining position dramatically improved due to the labour shortage. Many employees demanded loans from the fishing companies, creating a financial burden for some vessel owners.[5]

It was impossible to solve the problem of the labour shortage by simply increasing wages.[6] In 1978, the average annual catch by a Taiwanese fisher was three tons, and it remained the same in 1984. In other words, during those six years Taiwanese fishers did not improve their productivity as fish stocks declined and new coastal fishing boundaries were imposed.[7]

Fragile Financial Strength

Most Taiwanese vessel owners were financially weak; often their only assets were one or two fishing vessels. As a result, they depended heavily on loans from international fish dealers, banks and governments. They paid shipbuilders by instalments for new fishing vessels, and after the ships were complete they borrowed operating funds from fish dealers. To economize on personnel, most companies hired only a few essential office staff. The small administrative scale of these companies meant that most were not capable of negotiating with foreign courts or governments if a mishap occurred. Moreover, their limited financial capabilities prevented them from establishing their own communications networks.[8] Table 9.1 shows that in the early 1970s most distant-water longline fishing companies owned only two vessels, and some owned only one. This was also the case among Taiwan's trawler companies. The membership list of the Kaohsiung Fishing Boat Commercial Guild (KFBCG) shows that most members owned only one or two units of pair-trawlers. Table 9.2 shows that most distant-water fishing companies in Taiwan owned fewer than 500 tons of shipping. Significantly, the size of a distant-water fishing company in the 1980s was small as in the 1970s.

---

[5]*Fisheries Tribune*, No. 103 (1973), 4.

[6]Vessel owners did, however, loan them more money.

[7]Huang, "Economic Development," 9-12.

[8]*Fisheries Tribune*, No. 97 (1971), 9 and 12; and *A Report on Taiwan's Distant Water Longlining Fishing Industry* (Taipei, 1972), 73.

Table 9.1
Size of Taiwanese Distant-Water Longline Fishing Companies, 1970-1971

Year	Number of Companies	Number of Vessels	Average Number of Vessels per Company
1970	231	434	1.9
1971	260	462	1.8

Source: *A Report on Taiwan's Distant Water Longlining Fishing Industry* (Taipei, 1972), 66-67.

Table 9.2
Size of Taiwanese Distant-Water Fishing Companies, 1980
(Longliners and Pair-Trawlers included)

Total Tons per Company	Number of Companies	Percent
Less than 500 tons	1227	89.0%
500-1000 tons	129	9.5
More than 1000 tons	21	1.5

Source: Huang Sianchih, "Economic Development and the Structural Adjustment of the Production of Taiwan's Fisheries" (unpublished paper presented to the Joint Annual Meeting of ROC Agricultural Groups, Taiwan, 1984), 9-12.

Depletion of Marine Resources

A large number of pair-trawlers in postwar Taiwan were engaged in the demersal fisheries. They harvested fish in Southeast Asia and beyond until the resource was virtually depleted. Natural resource conservation and the rehabilitation of seabed ecosystems were not considered significant issues by either the Taiwanese government or the vessel owners. The problem of over-fishing had emerged initially in the East China Sea and the waters off Keelung in the early 1950s, and then in the South China Sea in the late 1960s. The trawler companies had two damaging solutions to the problem. First, they used small-mesh nets so they could harvest undersized fish. Second, they built larger vessels to exploit more remote fishing grounds off Indonesia and northern Australia. The former approach was both short-sighted and reckless. To harvest younger fish would only further accelerate the depletion of marine resources and eventually lead to a decline in the number of mature fish.[9] Under the circumstances, the rehabilitation of marine ecosystems in Asian waters had become nearly impossible because they had been severely damaged, in no small measure by Taiwan's large-scale, industrialized trawler fleets. The Taiwanese

[9] *Fisheries Tribune*, No. 21 (1963), 2.

fishing industry faced a grave crisis as catches began to plummet after 1972. Figure 1 depicts this rapid decline in catches in the waters off East and Southeast Asia in the 1970s caused by declining fish populations.

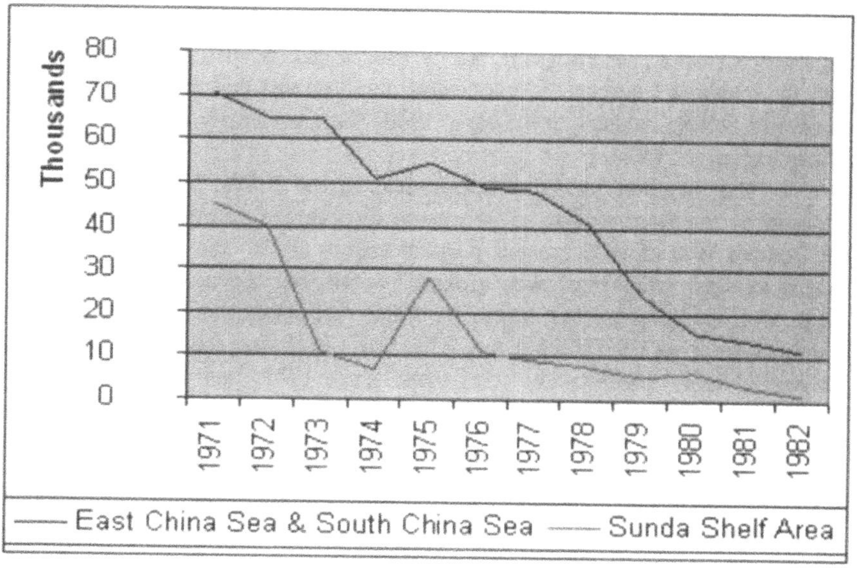

Figure 9.1: Catches off East and Southeast Asia, 1971-1982 ('000 metric tons)

Source: National Taiwan University, Institute of Oceanography, "Annual Report of Catch Statistics by Area on Taiwan Demersal Fish Fisheries," 1971-1982.

The rapid depletion of marine resources had a significant effect on the productivity of Taiwan's fishing industry. From 1953 to 1973, the average annual productivity growth of Taiwan's fishing industry was eight percent, but from 1973 to 1978 it dropped to only three percent, and from 1979 to 1983 to a meagre 0.034 percent. These figures demonstrate that after 1979 the industry scarcely made any productivity gains.[10]

Energy Crisis of 1973

As an island nation which lacked its own petroleum resources, Taiwan's fuel prices were comparatively high. The price of vessel fuel increased frequently depending upon global supply and demand. The price of diesel fuel increased from US $51.14 per kilolitre in 1955 to US $60 by 1962, a rise of seventeen

[10]Huang, "Economic Development," 9-12.

percent. The price of *Yakitama* engine fuel increased from US $24.19 per kilolitre in 1955 to US $37 in 1962, a price hike of fifty-two percent.[11]

Fuel prices rose every year between 1955 and 1962, but the price of fish products produced by the distant-water fishing industry did not increase as rapidly. In 1955 the average price of fish products from the distant-water fisheries was NT $4.84 per kilogram, but by 1962 it had only risen to NT $5.07, a growth of only 4.7 percent.[12] This clearly demonstrates that returns in the distant-water fishing industry in the early 1960s were not nearly as high as in the 1940s and early 1950s.

The situation radically deteriorated in the 1970s. The oil embargo imposed by the Organisation of Petroleum Exporting Countries (OPEC) after the October War of 1973 created a global energy crisis. The price of fuel for fishing vessels, which had been around US $40 per ton at overseas fishing ports, soon soared to historic highs. By 1984, fuel prices had increased in only a single decade to US $350 per ton. This had a profound economic impact on the operations of distant-water companies. Since 1973, fuel prices in Taiwan had also soared. Before the energy crisis diesel fuel was about US $57.80 per kilolitre, while fuel for *Yakitama* engines cost about US $35.60 per kilolitre. By 1984, the price of diesel fuel had increased nearly four-fold to US $223.37, and fuel for *Yakitama* engines had risen to US $156.48 per kilolitre. The sharp escalation in fuel prices created major financial problems for offshore longliners and distant-water pair-trawlers using Kaohsiung as a home port.[13]

Chronic Failure of International Fisheries Cooperation Arrangements

The establishment of cooperation arrangements in the international fisheries had a variety of advantages for Taiwanese fishing companies since they facilitated the effective exploitation of foreign fishing grounds; increased opportunities for domestic employment; added to the national wealth; strengthened foreign exchange; and encouraged the development of ancillary industries. Hence, both companies and the fisheries authorities believed that the future development of Taiwan's fishing industry would depend heavily on international coop-

---

[11] *Fisheries Tribune*, No. 23 (1963), 2. The question of why fuel prices in Taiwan increased so frequently can be explained by the fluctuating exchange rates between the NT $ and US $. See Paul C.H. Chiu, "Money and Financial Markets: The Domestic Perspective" in Gustav Ranis (ed.), *Taiwan: From Developing to Mature Economy* (Boulder, CO, 1992). In 1955 the exchange rate ($NT/$US) was 15.55:1; by 1962 it had reached 40:1.

[12] *Fisheries Tribune*, No. 23 (1963), 14.

[13] Huang, "Economic Development," 9; and *Fisheries Tribune*, No. 91 (1971), 4.

eration. Both groups tried hard to negotiate such arrangements with foreign countries. Indeed, as early as 1955 the China Fishing Company (CFC) was authorized by the Ministry of Economic Affairs to carry out this task on behalf of the government.[14]

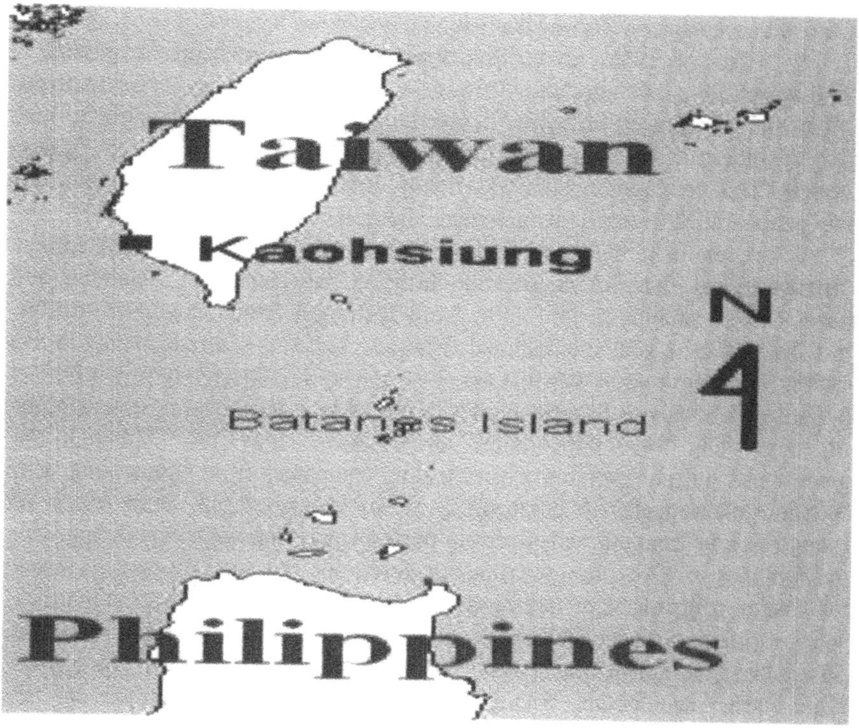

Figure 9.2: Kaohsiung and the Batanes Islands

Source: Courtesy of the author.

In the 1950s, the Philippines and Vietnam were the major countries in the region with which Taiwanese vessel owners wanted to cooperate. The waters off the two nations were considered by Kaohsiung's vessel owners to be the most important distant-water fishing grounds. The earliest attempt to negotiate a cooperative arrangement was in 1956 when the Jhongfei Fishing Company tried to make a private arrangement with a Filipino fishing firm. This plan, however, was never put into practice. At about the same time, the provincial government of the Batanes Islands in the northern Philippines (see figure 9.2) discussed the possibility of inviting Taiwanese vessels and fisheries technicians to explore the fishing grounds and marine resources off its shores.

[14]*Fisheries Tribune*, No. 10 (1962), 3, and No. 18 (1962), 2.

This plan also failed to come to fruition for three reasons. The first involved a boundary dispute between Taiwan and the Philippines that called into question the provenance of the fishing grounds. Second, the port facilities in the islands were inadequate to support a fleet of modern Taiwanese fishing vessels. More important, however, the Filipinos asked for large financial guarantees which were not acceptable to Taiwanese fishing companies.[15]

Also in 1956, some Sino-Vietnamese entrepreneurs expressed a willingness to cooperate with Taiwanese fishing companies. The fisheries authorities on Taiwan, however, doubted whether the ancillary industries and port facilities in Vietnam were good enough to service Taiwanese fishing vessels. This cooperation plan was finally aborted in August 1957 when the two parties failed to reach an agreement about its implementation.[16]

A report in the *China Fisheries News* in 1959 emphatically stated that "[i]nternational fisheries cooperation projects with Southeast Asian nations have achieved nothing so far." Why were Taiwanese fishing companies unable to work out successful international fisheries cooperation arrangements in the 1950s? The main reason was that the government did not actively support these initiatives. Taiwanese fisheries policy at the time gave the sole responsibility for negotiating these cooperation agreements to the CFC. Although the government did not prohibit private fishing companies from cooperating with foreign governments or companies, it never encouraged these kinds of initiatives. For example, when private fishing companies were negotiating with the Filipinos in 1956, the government offered no assistance. Instead, it raised two major concerns with the vessel owners. First, since Taiwan still had a serious dispute with the Philippines over the territorial boundary in the Bashi Strait, the government asked how the companies could disregard this serious issue during negotiations. Second, since at the time the Philippines did not have a proper set of laws governing fisheries cooperation, how could vessel owners possibly reach an agreement with the Filipinos?[17] It was obvious that the government was trying to discourage the vessel owners from undertaking any cooperation plan on a unilateral basis.

Furthermore, Taiwanese fisheries authorities never stipulated the relevant laws and regulations to which fishing companies were required to adhere under such circumstances. Vessel owners who intended to cooperate with foreign companies had no idea what legal procedures they should observe

---

[15]*China Fisheries News*, 19 and 26 November 1956 and 28 October and 11 November 1957.

[16]*Ibid.*, 4 March and 26 August 1957.

[17]*Ibid.*, 5 November 1956 and 28 December 1959.

from an international legal standpoint and with whom they should negotiate.[18] These two reasons help to explain why all prospective private fisheries cooperation plans were aborted in the 1950s.

On several occasions the CFC also attempted to negotiate cooperation projects with other countries. None of them, however, was successful. In 1961, the CFC and the government of South Vietnam planned to establish the Sino-Vietnamese Fishing Company. The CFC would provide fishing technicians, two vessels and employment opportunities, while the Vietnamese would open their fishing grounds and markets. Unfortunately, prior publicity about the proposed project, and the apparent richness of the marine resources, raised Vietnamese concerns. When the talks stalled, the Vietnamese navy began to detain Taiwanese fishing vessels more often than before. Relations between the two parties were soon strained.[19]

In the second half of 1961, the CFC carried out a fisheries cooperation project with a Thai fishing company. This project, however, turned out to be a huge failure. According to the agreement, the CFC was to provide two single-trawlers and employment opportunities, while the Thais would open their fishing grounds and pay the operating costs of the vessels, which were required to use Thailand as a supply base and to sell their catch in the Thai market. Profits would be shared equally. The failure of the agreement can be explained by three factors. First, Taiwanese trawlers harvested about 30,000 kilos of fish on their first trip, which could easily have been sold in Taiwan's domestic markets in less than a day. In Thailand, however, there was little demand for fish products, so Taiwanese fishers reluctantly discarded some of their catch. Second, the fish products harvested by the Taiwanese trawlers could only be sold in Bangkok due to Thailand's rather inefficient transportation system. Moreover, the ice supply in Thailand was inadequate to service the trawlers. Third, while in Taiwan trawler companies could earn between NT $150,000 and NT $200,000 from a 30,000-kilogram catch, in Thailand, they only received the equivalent of about NT $24,000. According to the contract, the trial period for this arrangement was to be six months, but the CFC wanted to terminate it after the first month. This move to abrogate the agreement provoked the Thais' anger. Two trawlers were quickly detained by the Thai government, and all the Taiwanese crew, including a manager from the CFC, were held in custody. With the help of the Taiwanese Embassy, the

---

[18]*Ibid.*, 5 September 1958.

[19]*Fisheries Tribune*, No. 6 (1961), 2, and No. 8 (1962), 8; "A Brief Introduction to State-Run Industries, 1961," in *The References of the Ministry of Economic Affairs* (Taipei, 1961), CCXXXII, 19; and *China Fisheries News*, 6 November 1961.

fishers and staff were eventually released, but the two single-trawlers were sold to local Chinese at cut-rate prices.[20]

In fact, the Thai government had every reason to detain the two Taiwanese vessels. The cooperation arrangement between the two parties required the prior approval of the Thai government, but the Thai fishing firm that cooperated with the CFC was a private company which did not cooperate with the government. The Taiwanese Embassy in Bangkok had already warned of the possible dangers of entering into such an arrangement, but the CFC insisted upon taking the risk.[21]

The international fisheries cooperation projects carried out by the CFC repeatedly failed, which provoked resentment among key figures in Taiwan's fishing industry. The public expected that the personnel of the CFC would be reorganized and that Chen Liang, its head, would be dismissed. His position, however, was never threatened. A plausible argument was made by the *Fisheries Tribune* in 1962. To cover up internal mismanagement, the CFC continually sought opportunities to cooperate with foreign fishing companies even when the projects were not feasible.[22]

The CFC was finally dismantled in 1965, and four years later the Kaohsiung fishing industry, with the help of the Fisheries Bureau, began negotiations with the Indonesian government. Taiwanese representatives put forward two requirements. First, the Indonesian government had to furnish supplies and medical care to Taiwanese fishers and vessels and guarantee their safety in Indonesian fishing ports and territorial waters. Second, as a form of remuneration the Indonesian government would receive ten percent of the profits generated from the marketing of the fish products. The former requirement was accepted by the Indonesians without reservation. Concerning the latter, the Indonesians wanted thirty percent of the profits. Alternatively, they suggested that the ownership of the Taiwanese vessels that participated in the fishery be transferred to Indonesian partners after a stipulated period of time. This demand was considered unreasonable by the Taiwanese vessel owners. In addition, different military commands in Indonesia had different views on the project. The situation soon became so complicated that the vessel owners were forced to give up.[23]

---

[20]*Fisheries Tribune*, No. 8 (1962), 8, No. 9 (1962), 3, and No. 11 (1962), 10-11. Each of the trawlers cost at least NT $3 million, but they were sold for NT $1.2 million.

[21]*Ibid.*, No. 11 (1962), 10-11.

[22]*Ibid.*, No. 8 (1962), 8, and No. 10 (1962), 3.

[23]*Ibid.*, No. 79 (1969), 9.

In mid-1974, the KFBCG once again tried to initiate a trial cooperation agreement with Indonesia. The plan now was to form a partnership with two Indonesian companies, the Jhongshih Company and the Koda Jaya Company. According to the draft proposal, Taiwanese vessel owners would have to offer job opportunities to Indonesian fishers. Further, each trawler would have to pay an annual fee of US $5000 to the Indonesian government through Jhongshih, which would then apply for a fishing licence, travel permit and official approval from the Indonesian fisheries authorities and the military, which was to guarantee the safety of the Taiwanese vessels. Also, Jhongshih would have to negotiate with the Indonesian government on behalf of its Taiwanese partners if a detainment occurred. Koda Jaya would have to make its telecommunication system available to the Taiwanese trawlers participating in this cooperation arrangement.[24]

But this proposed plan also miscarried due to a serious dispute between the Kaohsiung fishing companies and the Indonesian government. In order to exploit particular fishing grounds off Australia, in the 1970s a growing number of Taiwanese fishing vessels were passing through Indonesian waters. By mid-1974, in order to prevent their vessels from being detained, Taiwanese vessel owners began to purchase special travel permits for the right of passage from the Indonesian government through the Indonesian Visa Office in Taipei. Unfortunately, vessels were detained as often as before. Kaohsiung vessel owners felt that they were being deceived by the Indonesians and asked the Taiwanese government to intervene in the dispute. They also wanted to sue the Indonesian government, but this move was blocked by the Taiwanese government. It was estimated that eighty-three units of pair-trawlers in Kaohsiung had purchased travel permits from the Indonesian government for a sum equivalent to NT $14,324,990. Without the government's full support, the Kaohsiung fishing companies were forced to tolerate the fraud and deception of the Indonesian authorities.[25] The detainment incidents forced Guild members to reject any thoughts of cooperating with Jhongshih and Koda Java.

Throughout 1973 and 1974, the KFBCG also contemplated fisheries cooperation arrangements with South Vietnam. In the early 1970s, Taiwanese vessels were fishing in the waters off Indonesia and Australia, but the grounds off South Vietnam were considered too old and unproductive by Kaohsiung's

---

[24]KFBCG, "10th Meeting Records of the 5th Cohort Directors," 6 July 1974; and "The Memorandum of Fisheries Cooperation...from 1st Temporary Conference Records [Agenda] of the 5th Cohort Member Representatives," 20 August 1974.

[25] *Ibid.*, "10th Meeting Records of the 5th Cohort Directors;" "8th Temporary Combined Meeting Records of the 5th Cohort Directors and Supervisors," 18 October 1974; "12th Meeting Records of the 5th Cohort Directors," 16 November 1974; "13th Combined Meeting Records of the 5th Cohort Directors and Supervisors," 24 January 1975; and "14th Meeting Records of the 5th Cohort Directors," 7 April 1975.

trawler owners. The main reason that the KFBCG wanted to work with the Vietnamese was to search for new fishing grounds for its old pair-trawlers, some of which (mainly those between 120 to 150 tons) were still in active service. They were too small to fish in the waters off northern Australia but too large to work in the Taiwan Strait. Hence, in 1974 the Guild sent a delegation to investigate the possibilities. Once again, however, poor port facilities and marketing systems compelled them to drop the idea.[26]

In 1975, another fleet of fishing vessels from Kaohsiung tried to establish a cooperative agreement with Indonesia. This effort turned out to be the largest setback in the history of Kaohsiung's fishing industry, as the vessel owners had no idea what legal steps to take in case of a mishap or how to deal effectively with the Indonesian government.

Before leaving Kaohsiung for the eastern Indonesian fishing grounds, they failed to inform the Indonesian Visa Office in Taipei of their plans. Moreover, they were not aware that they also had to apply for a travel permit and official approval from both the Indonesian Department of Defense and Security and the Department of Agriculture. When asked to show the certificates of registration of the vessels' nationality, the Taiwanese fishers were unable to provide the master copies. As a result, the officials accused them of illegally entering Indonesian territorial waters. But the Indonesians did not want to bring this case to court and quietly asked the fishing masters to adopt a more conciliatory stance. This meant that Taiwanese fishers would have to pay the penalty of concession fees in order for the Indonesian government to drop the matter.[27]

The vessel owners, however, rejected this offer. To salvage the precarious situation, they hoped to cooperate with another Indonesian fishing enterprise and reapplied for permission from the relevant Indonesian authorities. To further increase their chances of success, the proportion of capital invested by the Taiwanese and Indonesians was now significantly readjusted; Taiwanese funds accounted for forty-nine percent and the Indonesians were given a commanding fifty-one percent. Through this strategy they obtained the necessary permission from the Investment Coordination Board and the Department of Agriculture. The Indonesian military, however, refused to ratify the business licence for the proposed project.[28]

---

[26]*Ibid.*, "5th Meeting Records of the 5th Cohort Directors," 13 August 1973; "7th Temporary Meeting Records of the 5th Cohort Directors," 1 August 1974; and Li Jijhao, interview, 22 March 2002.

[27]Fu Kuncheng, *A Study of the Foreign Courts' Decisions Regarding the Illegal Entry of Taiwan's Fishing Vessels* (2 vols., Taipei, 1990), I, 510.

[28]*Ibid.*, I, 510-518.

Instead, for reasons that remain unclear, the Indonesian lawyers reversed their position on an out-of-court settlement and decided to resubmit the case to the courts. The representatives of the Indonesian navy and the head of the Samarinda Port Authority became star witnesses for the prosecution. The court ruled that the Taiwanese fishers had illegally violated Indonesia's sovereignty and ordered them to pay NT $4 million in damages. Although this was a huge sum in 1975, the rapacious Indonesian military officials were still not satisfied. They asked the prosecutors to take further action by accusing the Taiwanese fishers of smuggling. This would give Indonesian officials a legitimate reason to confiscate the nine well-equipped Taiwanese fishing vessels. When this occurred, the Taiwanese government simply reiterated its official position that before entering into any fisheries cooperation arrangements with foreign nations, vessel owners should first consult the government.[29]

In February 1976 two Taiwanese fishing companies, Rihsheng and Rihlong, tried to enter into a cooperative agreement with a Malaysian company, the Dayalian Chanye Company. Dayalian would apply for entry and fishing permits for both Taiwanese companies in exchange for eighteen percent of the profits from each fishing voyage. In May of the same year, Taipei Fangjheng Trade Company also tried to set up a fishing company in Jakarta. According to the proposed contract, the Indonesians would receive five percent of the profits from each fishing voyage and ownership of the fishing vessels after fifteen years.[30] Both these plans also miscarried.

Why did the international fisheries cooperation projects that the Taiwanese fishing industry attempted to carry out always fail? Obviously, vessel owners did not have a sufficient understanding of local practices. As a result, frauds and detainments frequently occurred when Taiwanese fishing vessels operated in the waters of other coastal nations. To solve this problem, in 1976 the Fisheries Bureau of Taiwan planned to set up a civilian organization to collect overseas fisheries information and to help vessel owners negotiate with the authorities of other nations to establish cooperative projects. This proposed organization would be a corporate body which could raise the necessary funds from private distant-water fishing companies; any shortage would be covered through government subsidies.[31]

Facilitating a pattern of fisheries cooperation agreements was one of the most important strategies adopted by Taiwan's fishing industry to cope with the serious challenges presented by the implementation of the 200-mile

---

[29]*Ibid.*, I, 510; and KFBCG, "2nd Meeting Records [Agenda] of the 6th Cohort of Temporary Directors," 22 June 1976.

[30]*Chinese Fisheries Weekly*, 16 February and 10 May 1976.

[31]*Ibid.*, 13 December 1976 and 18 July and 8 August 1977.

EEZs. Yet they were never really successful no matter how hard they tried. Compared to other fishing powers, Taiwan had far more difficulty in carrying out such arrangements for several reasons. First, Taiwan was (and still is) isolated in the international arena due to Chinese pressure and influence. As a result, the government could never fully support the implementation of a fisheries cooperation programme or properly protect the offshore business interests of Taiwanese fishing companies.[32] Second, the laws relating to fisheries cooperation were not favourable towards Taiwanese fishing vessels undertaking international fisheries cooperation agreements with foreign countries. In an early period, the government of Taiwan had encouraged Western nations and Japan to invest on the island. Yet it never encouraged its fishing enterprises to invest in Southeast Asian nations because of the financial risks involved. Some of the significant issues included whether the vessel owners could operate a fishing company profitably in conjunction with foreign enterprises and whether the fish products from such jointly-owned fishing companies could actually be sold in Taiwan. For more than a decade the government never bothered to clarify its position on these issues, much to the chagrin of the vessel owners. This ambiguous attitude discouraged those in the fishing industry who were interested in establishing international fisheries cooperation agreements in Southeast Asia during the 1960s and 1970s.[33]

Impact of the Implementation of 200-mile EEZ

It was widely recognized by the late 1970s that if a 200-mile EEZ became common, the fishing grounds where distant-water fishers could freely harvest would suddenly be reduced by two-thirds, and that eighty percent of the world's fish resources would be "fenced in" and monopolized by coastal nations. Before the United States announced its 200-mile EEZ, people in Taiwan had already begun to predict the possible impact such a move would have on the distant-water fishery. First, the US would no longer share its fish with foreign fleets as before. Second, other island or coastal nations would most likely follow in the footsteps of the US and claim 200-mile EEZs. Hence, global marine resources would eventually be divided among coastal states according to the limits of their maritime boundaries.[34]

In the 1970s, not many Taiwanese vessels worked in American coastal waters, so its fishing industry felt it could survive even when the US actually established its EEZ. What created panic in Taiwan's fishing industry

---

[32]*Ibid.*, 19 September 1977.

[33]*Ibid.*

[34]*Ibid.*, 22 December 1975 and 16 August 1976; and Wang and Liou, *Problems of Taiwan's Distant Water Fisheries*, 95.

was the feeling that other nations would soon follow the American example and announce their own 200-mile EEZs. The fears of the Taiwanese vessel owners proved to be correct. When the implementation of the 200-mile EEZ was still in the planning phase, the Soviet Union did not lend its support, but it made a *volte-face* and also claimed a 200-mile EEZ on 1 March 1977, the same day that the US claimed its own 200-mile zone.[35]

Taiwan was not the only distant-water fishing nation to suffer from the rapid implementation of the 200-mile EEZs being established by countries around the world. Its neighbour, Japan, was also hurt badly by such measures. The northern waters off Soviet East Asia were the traditional fishing grounds of Japanese fishers; in 1976 they harvested 1.1 million tons of fish in the Bering Sea alone, accounting for one-tenth of the total yield of the Japanese fishing industry. To minimize the possible economic impact of the 200-mile EEZ, Japan planned to reduce the scale of its distant-water fishing fleet by a third. Taiwan's economic situation and prospects were not significantly better than those of Japan. Taiwan's fishing industry was concerned to see how the southern hemisphere nations would respond since the waters off Indonesia and Australia were important fishing grounds for its pair-trawlers, and the waters off some island nations in the South Pacific were traditional fishing territories for its distant-water longliners. The implementation of the 200-mile EEZ cast a dark shadow over the future of the Taiwanese distant-water fishing industry.[36]

The member states of the Pacific Island Countries and Territories, including French Polynesia, Papua New Guinea, Fiji and others, all planned to claim 200-mile EEZs. Ratu Sir Kamisese Mara, the Fijian prime minister, stated publicly that the waters for 200 nautical miles off the shore of Fiji's outermost islands would comprise his country's EEZ. The independent state of Western Samoa also decided to claim a 200-mile EEZ by August 1977. This news badly frightened vessel owners in Kaohsiung because these waters were crucial fishing territories for Kaohsiung's longline industry.[37]

On 12 May 1977, two years after the end of the war with the United States, Vietnam claimed a 200-mile EEZ and began to harass Taiwanese pair-trawlers in the South China Sea. The Philippines also announced a 200-mile EEZ on 11 June 1978, unilaterally enclosing its fishing grounds and presenting the Kaohsiung longliners with another difficult problem. Hence, the rapid implementation of a series of 200-mile EEZs in the late 1970s had a tremendous impact on the management of marine natural resources and the

---

[35]*Ibid.*, 16 February 1976; and Wang and Liou, *Problems of Taiwan's Distant Water Fisheries*, 95.

[36]*Chinese Fisheries Weekly*, 22 December 1975, 16 August 1976 and 29 August 1977.

[37]*Ibid.*, 29 August 1977.

exploitation of distant-water fisheries. The traditional concepts of the "commons" and "freedom of the seas" were now severely challenged. Vessel owners in Kaohsiung faced a terrible threat to their industry and way of life.[38]

To overcome the trauma of the introduction of the 200-mile EEZ, the fisheries authorities, marine scientists and representatives of the fishing industry convened a special conference in Taipei in 1977 to discuss possible solutions. They reached several key conclusions. First, the distant-water trawling industry would be the first to bear the full brunt of the EEZs as they came into effect. Second, due to Taiwan's continued diplomatic isolation, carrying out effective fisheries cooperation arrangements with other nations would continue to be difficult. Third, the future of Taiwan's distant-water fishing industry lay in the development of new fisheries and technologies, like purse-seining and deep-sea trawling. Both of these could effectively operate on the high seas and consequently avoid the direct impact of the 200-mile EEZs.[39]

The implementation of the new regime created panic among Taiwanese fishing companies which, at the beginning of the 1970s, had just built new trawler fleets to exploit the distant fishing grounds off northern Australia. Government officials and fisheries scholars tried to calm the vessel owners. Yang Jicyuan, the Vice-Minister of Economic Affairs, stated publicly that the implementation of the 200-mile EEZ did not spell the end of Taiwan's distant-water fisheries. Taiwanese fishing fleets still could operate in the EEZs of other nations through international cooperation arrangements, especially in the waters of affluent nations with abundant land resources like the US, Canada, Australia and New Zealand, because citizens in those countries did not necessarily need to make a living from the sea. Also, smaller, less-developed nations without adequate fishing technology and fleets to exploit their own rich marine resources might allow Taiwanese vessels to fish in their EEZs because the access fees that the fishing companies would pay would be a major source of revenue.[40]

---

[38]"Statement on the Territorial Sea, the Contiguous Zone, the Exclusive Economic Zone and the Continental Shelf of 12 May 1977," *Maritime Space: Maritime Zones and Maritime Delimitation*, http://www.un.org/Depts/los/LEGISLATION ANDTREATIES/, accessed 24 October 2005; Wang and Liou, *Problems of Taiwan's Distant Water Fisheries*, 111; "Presidential Degree No. 1599 of 11 June 1978 Establishing an Exclusive Economic Zone and For Other Purposes," Lawphil Project, Arellano Law Foundation, http://www.lawphil.net/, accessed 23 March 2006; and *Chinese Fisheries Weekly*, 19 December 1977.

[39]*Chinese Fisheries Weekly*, 24 January 1977.

[40]Lin Changren, interview, 17 May 2002; Gu Tingfang, interview, 7 April 2002; *Chinese Fisheries Weekly*, 19 December 1977; and "The Pacific Tuna Fishery," *ESCAP Virtual Conference*, http://www.unescap.org/DRPAD/VC/conference/ex_pi_17_ptf.htm, accessed 3 July 2006.

Yang also noted that there were many different high-value fish species to be found in the open seas. He encouraged Taiwanese fishing companies to develop other fisheries rather than simply sticking to trawling and longline fishing. In addition, the fisheries authorities had been thinking about the potential of exploiting the rich shrimp resources in the waters off Antarctica. *Haigong Hao*, the large Taiwanese experimental vessel that accommodated both marine scientists and fishers, undertook its pioneering voyage to Antarctica at the beginning of 1977. The results of the fishing experiments were very encouraging. The scientists on board *Haigong Hao* described this voyage as "a breakthrough in Taiwanese fishing history" and optimistically predicted that the rich shrimp resources off Antarctica and could be exploited on a commercial basis.[41]

There was a major obstacle to the development of these new southern fisheries, however: all of them required huge financial investments. A unit of large purse-seiners cost approximately NT $70 million (US $1.84 million), and a deep-sea trawler cost about NT $80 million (US $2.1 million). Due to the small size of most Taiwanese fishing companies, few could afford the expense of such a venture even though the government promised loans. Why did it cost so much to run these new fisheries? When a deep-sea trawler was operating, its net had to go down to a depth of at least 400 metres, which required a large fishing vessel with a powerful engine. The organization of purse-seining also required an entirely new fleet. Further, harvesting shrimp in the waters off Antarctica required vessels of at least 1000 tons and accompanying factory ships, and few Taiwanese fishing companies had the capital to build or operate such a huge enterprise. Moreover, most Taiwanese vessel owners knew next to nothing about the grounds for purse-seine and deep-sea trawling or the techniques necessary to exploit them, and the experience and quality of the Taiwanese fishers in the 1970s was not conducive to learning these new fishing methods rapidly. In the mid-1970s, the minutes of a meeting of the KFBCG reported that "[s]o far, not many vessel owners are interested in the investment in this fishery."[42]

In addition to investing in new fisheries, another possible way to overcome the hurdle of the 200-mile EEZs was to undertake a series of international fisheries cooperation arrangements. If these were successfully put in place, Taiwanese pair-trawlers and longliners could continue to fish on their current grounds. As we have seen, however, from the mid-1940s to the late 1970s the Taiwanese government and the private fishing companies were never able to implement these projects successfully. But Kaohsiung's vessel owners

---

[41]*Chinese Fisheries Weekly*, 1 March, 25 April and 19 December 1977.

[42]*Ibid.*, 24 January, 1 March and 30 May 1977.

still believed that these arrangements comprised the correct strategy and were seriously thinking about raising funds to lobby foreign governments.[43]

As late as 1976 and 1977, the KFBCG was still discussing possible arrangements with Indonesia, New Zealand and Australia. The process, however, required lengthy negotiations. The Taiwanese fisheries authorities tried to help, but Taiwan's diplomatic isolation prevented it from negotiating effectively with foreign nations, since they could only speak with foreign officials privately due to the constant pressure being exerted by China.[44]

The exploitation of living marine resources has to a certain extent always been an extension of international politics. Fishing rights could readily be used as bargaining chips in international political and economic negotiations. The USSR's claim to a 200-mile EEZ in 1977, in the opinion of the Taiwanese government, was a warning to Japan not to align itself too closely with China. Robert Muldoon, the Prime Minister of New Zealand, stated that Japan would have to open its markets for New Zealand's agricultural products if it wanted to exploit the marine resources inside New Zealand's EEZ. Muldoon's statement made Taiwan's fishing industry ponder the fact that, as a politically isolated nation, Taiwan had little to offer to other countries in exchange for a fisheries cooperation agreement. A fisheries cooperation arrangement between Taiwan and Australia finally began in November 1979. Some Taiwanese vessel owners also managed in the same year to cooperate informally with the Indonesians, but this was stopped by the Indonesian government in 1982. While a fisheries cooperation arrangement with India began in 1982, not many Taiwanese pair-trawlers liked working on that remote fishing ground.[45]

It is not hard to imagine that between 1977, when the 200-mile EEZs were commonly claimed, and 1979, when Taiwan began to implement fisheries cooperation agreements with other countries, Taiwanese vessel owners were extremely anxious. They still could not feel secure several years later, although they had successfully carried out some forms of cooperation with Australia and Indonesia, because in order to maintain such relations, the Taiwanese distant-water fishing fleets had to be closely supervised by other

---

[43]KFBCG, "19th Combined Meeting Records of the 5th Cohort Directors and Supervisors," 22 January 1976; and "1st Temporary Conference Records of the 6th Cohort Member Representatives," 2 July 1976.

[44]*Ibid.*, "6th Meeting Records (Agenda) of the 6th Cohort Directors," 7 December 1976; "5th Meeting Records (Agenda) of the 6th Cohort Directors," 19 October 1976; "2nd Temporary Meeting Records (Agenda) of the 6th Cohort Directors," 22 June 1976; and *Chinese Fisheries Weekly*, 25 October 1976.

[45]*Chinese Fisheries Weekly*, 25 October and 5 December 1977; and "The Running of Trawl Fishing" (unpublished KFBCG internal document, n.d.), 1.

nations. Fishers had to provide honest reports to local fisheries officials about the species they harvested, the volume of their catch and the methods they used. Hence, fishing companies had to learn the fisheries regulations of other nations, and fishers had to learn foreign languages to communicate effectively with the law enforcement officials of neighbouring coastal states.[46]

Several things can be concluded from all this about the significance of the imposition of the EEZs. First, the development of Taiwan's distant-water fishing industry was affected by several factors, including a labour shortage, fragile financial conditions, the depletion of regional marine resources and a continual failure to establish international fisheries cooperation agreements. While these issues on the surface appeared to be separate, in fact they were all closely interlinked. Running deep-sea trawlers and purse-seiners required high-quality crews, but their recruitment was almost impossible considering the labour shortage. Building new boats and exploring remote fishing grounds required a large amount of capital, but most Taiwanese companies lacked the necessary resources. The oil crisis pushed fuel prices to historic highs, which placed the fishing industry in a tight economic squeeze.

The exploitation of more remote fishing grounds had always been used by Taiwanese vessel owners as a solution to the problem of marine resource depletion. For decades, a large number of Taiwanese pair-trawlers had harvested fish in distant waters. The rehabilitation of seabed ecosystems and the conservation of living marine resources had never really attracted the serious attention of either the Taiwanese fisheries authorities or the fishing companies whose modern, industrialized fleets had consecutively damaged fishing grounds in the East China Sea, Taiwan Strait, South China Sea and the waters off Vietnam, Indonesia and Australia. Taiwanese pair-trawlers had scoured these distant fishing grounds like swarms of locusts. It is fair to say that the postwar development of the Taiwanese trawling industry contained the seeds of its own destruction. Pair-trawler owners became victims of their own ecologically rapacious deeds in the 1970 and 1980s by turning some distant-water fisheries into "deserts." Like pair-trawlers, longliners needed to work in the waters of other nations. Every year, numerous small, but well-equipped, longliners from Kaohsiung and other ports in southern Taiwan fished off the Philippines. Fishing legally and safely in those waters required successful fisheries cooperation agreements, but before 1977 no such project had been successful. Even at the beginning of the twenty-first century, fisheries conflicts between the Taiwanese and Filipinos continue to occur.

The implementation of the 200-mile EEZ was a good thing from a regional conservation perspective because it effectively prevented Taiwanese trawlers from further exploiting the waters of other nations without their consent. But the implementation of the 200-mile EEZs also sounded the death

[46]*Chinese Fisheries Weekly*, 14 November 1977.

knell for Taiwan's trawling industry. Although Kaohsiung's trawl fishery survived through international fisheries cooperation arrangements with Australia, Indonesia and other nations for several more years, its demise occurred soon after these major cooperative projects ceased.

Conclusion

The development of Taiwan's fishing industry off Southeast Asia in the twentieth century can be described and analyzed at the state, industry and community levels. Viewed from the perspective of the state, it is clear that the government's policies and regulatory practices played significant roles in the birth and development of the industry since the colonial period. To avoid the growing problem of over-fishing in the East China Sea, the colonial authorities encouraged fleets to expand their fishing southward in the 1930s. At the same time, the Japanese military encouraged them to explore the hitherto unknown fishing grounds in the *Nanyo*, expecting fishers to collect strategic and nautical information for the navy. Hence, these fleets worked as pioneers for the southern expansion of the Japanese Empire.

As a key transport hub between East and Southeast Asia, Takao (Kaohsiung) was always regarded by the Japanese as a crucial fishing base on the empire's southern frontier, and they made a major, capital-intensive effort to develop the fishery. Modern fishing techniques were introduced from Japan, well-organized management systems were established and large-scale investigations into marine resources off Southeast Asia were undertaken. Most important, modern, comprehensive port facilities were constructed in Takao. Both domestic and international fish markets and the construction of efficient transport networks contributed greatly to the development of Takao's fishing industry. All these state-driven infrastructure projects laid a strong foundation for the rebirth of the fishing industry in postwar Taiwan.

Taiwan's fishing industry was almost totally destroyed during World War II, but it was soon revived as a direct result of the postwar government's intervention and financial assistance from the international community, especially the United States. The initial motivation for rebuilding Taiwan's fishing industry was to meet postwar consumption demands.

Fisheries development policies in postwar Taiwan had two key goals: to increase fisheries production and to build modern, well-equipped vessels with foreign financial assistance. These guiding principles served as a template for the growth of Taiwan's fishing industry. Meanwhile, large-scale financial assistance from the United Nations Relief and Rehabilitation Agency, the US government and the World Bank provided major external support that not only enabled the Taiwanese to rebuild their shattered fishing industry but also helped to build the confidence to establish a global-scale fishing industry.

The fisheries policies in postwar Taiwan did not deviate markedly from the earlier course set by the Japanese. Fisheries-related infrastructure continued to be constantly upgraded, and new fishing techniques were continually introduced. The increase in vessel tonnage and the modernization of fishing equipment were also still deemed essential, and the exploitation of Southeast Asia's marine resources was still considered the fleet's most important task. In the prewar era, fishing vessels had been encouraged to fish further southward, and in the early postwar years they were encouraged once again to exploit marine resources on the various fishing grounds off Southeast Asia. The motivation was both political and environmental. Besides compensating for the dwindling marine resources in the East China Sea by the 1950s, Taiwanese fishers also wanted to avoid possible military harassment by the Chinese communists, especially when the trawling grounds in the East China Sea shrank dramatically after Taiwan withdrew its garrison from Dachen Island in 1954.

Because the postwar government, like the Japanese colonial administration, believed that Kaohsiung's geographical location was crucial to the development of Southeast Asia's fishing grounds, it also put a huge effort into the reconstruction of Kaohsiung Fishing Port. The rebuilding and operation of Gushan Fishing Port not only encouraged Kaohsiung's longliner owners to re-exploit the marine resources off Southeast Asia but also attracted Keelung's trawling companies to relocate to Kaohsiung. These developments were followed by the construction of Cianjhen Fishing Port in 1967, which proved a milestone in Taiwan's maritime history. Ice-manufacturing plants, cold storage facilities, fish-processing factories and other ancillary industries now congregated in Kaohsiung. Fisheries-related agencies and institutes, including the Weather Station, Kaohsiung Training Centre, Fisheries Research Institute and Taiwan Area Fishery Radio Station, were all located in the port area. Armed with these modern facilities, Kaohsiung's fishing fleets became quite active on the fishing grounds off Southeast Asia in a comparatively brief period of time. Taiwan's distant-water fishing fleets now began to fish beyond the region and started to pioneer a global-scale fishery.

An examination of the historical scope and magnitude of Taiwan's fishing industry can be narrowed from the state to the industry level. When Taiwanese fishing vessels first entered Southeast Asian waters, Kaohsiung's fishing industry was also busy negotiating and dealing with the government, military, politicians and onshore ancillary industries.

Besides moulding the fishing industry into a national enterprise through policies and administrative intervention, the postwar fisheries authorities also controlled and regulated it because the potential mobility of fishers was viewed as a possible threat to Taiwan's political and social stability. Furthermore, the fishery was considered a reliable source of food for the military. This explains why the fisheries authorities, including the Taiwan Fisheries Production Committee, the Fisheries Administration Agency (renamed the

Fisheries Bureau in 1965) and the state-run fishing company (the China Fishing Company, or CFC) were dominated by people with military backgrounds.

The onshore activities of Kaohsiung's fishing companies were also under government surveillance. A representative of the Taiwan Garrison Command was permanently given membership in the Kaohsiung Fishing Boat Commercial Guild (KFBCG). Hence, representatives of the *Kuomintang* (KMT) and the police were always present whenever vessel owners met. In addition, the military in the early postwar years adopted vigorous measures to mobilize fishing fleets to assist in military actions against the People's Republic of China, something which placed a tremendous operational burden on the fishing companies.

Relations with the government and military amounted to more than direct interference and control. Facing intervention from the military and the government, the fishing industry seldom complained. Instead, it tried to establish good working relationships with both. This was partly because the trawler fleets working off China and in the South China Sea needed naval escorts, and partly because the fishing companies recognized that the military provided an important market, especially when domestic demand was depressed. Consequently, the burgeoning fishing industry in Kaohsiung seized every opportunity to demonstrate its loyalty and support to the KMT and the military.

The Kaohsiung fishing industry also asked for help from parliamentarians who had emigrated from mainland China, but they rarely received a positive response. While vessel owners also tried to get involved in local politics, they encountered a number of setbacks in the early days of the "White Terror" (1945-1987), when their spokesmen were ruthlessly killed, imprisoned on trumped up charges or driven out of politics. The national-level parliaments were basically controlled by mainland Chinese until the early 1970s. Hence, it was almost impossible for Taiwan's fishing industry to elect its own representatives in the national parliamentary system. The political situation did not improve until the elections of supplementary parliamentary members were held in 1972. The process to gain a direct political voice for the Kaohsiung fishing industry had taken almost three decades to accomplish.

To ensure that the long-distance fishing fleets working in Southeast Asian waters had adequate supply and support, Kaohsiung's fishing companies spent considerable time and effort dealing with ancillary industries, especially ice manufacturers, shipbuilders, trucking companies, equipment suppliers and security guards. Some of these ancillary industries were well-organized and financially strong, while others were merely small, retail businesses run by poor tradesmen and labourers. Nevertheless, all presented challenges to the progress and development of Kaohsiung's fishing industry. Sometimes fishing companies took joint action to fight against the monopoly practices and collusion of the various ancillary industries, while at other times they asked the authorities for assistance. Unfortunately, the government did not always side with the fishing industry in these disputes. Moreover, when facing challenges from the ancillary industries, the fishing companies were often divided.

The scope of the history of Taiwan's fishing industry in Southeast Asia can be further narrowed from the industry level to the actual fishing communities. The impoverished conditions and limited livelihoods in Penghu and Siao Liouciou, as well as the related problem of over-fishing in the East China Sea, drove numerous Penghuan and Siao Liouciou fishers, and a large number of trawler owners from Keelung, to Kaohsiung at different points in time. They took advantage of the modern port facilities to exploit the more distant marine resources off Southeast Asia, which also changed the ethnic character and composition of Kaohsiung's fishing circles and port life.

The selection of the type of fishing method for the long-distance industry was strongly influenced by ethnic factors. Most local vessel owners restarted their fishing businesses in postwar Kaohsiung by operating longline fishing companies, but quite a few diversified into trawling. While the fishing grounds off Southeast Asia were of primary importance, they also invested considerable effort in exploring more distant seas. Most mainland Chinese started their fishing careers in Taiwan by operating trawler companies, and they always stuck with this. While they invested in other fishing activities, the scale of these other enterprises could not compare with their trawling endeavours. The Penghuans comprised the main source of labour in the distant-water fisheries, especially on the longliners. They generally served as crew, and only a few managed to launch their own fishing companies. All the Siao Lioucious adhered solely to offshore longline fishing. They were the most conservative ethnic group in Kaohsiung's fishing industry. They fished on the grounds that the Penghuans left behind in Southeast Asia as they entered new fisheries. They shunned other ethnic groups, which limited their potential for participating in the emerging global-scale fisheries.

The vessel owners' attitude towards the government and military surveillance was affected by ethnic factors as well. When the military and the government mobilized their fishing fleets, mainland Chinese vessel owners did not complain too loudly. Instead, they demonstrated their patriotism and offered material support to the military because they had only recently escaped from China with the KMT; a foreboding sense of political crisis between Taiwan and China encouraged them to support the authorities and the military. Further, most of their pair-trawler fleets fished in the waters off China and Vietnam and required the regular protection of the navy. Finally, under Chiang's regime, the people who dominated the military were also mainland Chinese. Thus, the trawler owners felt far more comfortable dealing with these groups due to their similar ethnic backgrounds and political affinities.

When the 200-mile Exclusive Economic Zones (EEZs) came into force in the late 1970s, the ethnic groups applied different strategies to deal with this major challenge. Local Taiwanese vessel owners tried to arrange international fisheries cooperation, while others went back to distant-water longline fishing or adopted new techniques, such as gill-net or purse-seine fishing. When the golden age of trawler fishing ended in Southeast Asia, a large number of mainland Chinese vessel owners chose to leave the Kaohsiung fishing industry. This prudent decision helped them to conserve their wealth, but it also effectively prevented them from participating in the further expansion of Taiwan's global-scale fisheries in the 1980s that was spearheaded mainly by the longline fisheries.

Numerous fishers from Penghu, Siao Liouciou and other places in Taiwan migrated to postwar Kaohsiung, which ultimately limited employment opportunities in the fishing industry. Fishing masters preferred to employ relatives or those who were prepared to offer presents or bribes. This traditional custom further highlighted the ethnic and social boundaries between different groups of fishers. Most young men started their fishing careers as "cook's boys" or engineman assistants. But the responsibilities of being a cook or engineman assistant could be difficult and sometimes tragic. They were the busiest people on distant-water or offshore longliners. Young fishers often were physically abused and financially exploited by more senior crew. Sometimes these novices served as scapegoats on whom other fishers could vent their anger. Senior crew, like the fishing master, engineman and radioman, also found life at sea hard, as having the same ethnic background did not necessarily guarantee that interpersonal relations would always be harmonious. They had to take certain precautions against their fellow crew members. Sometimes they plotted against each other for personal reasons. Away from home, lonely young fishers missed their parents, and senior fishers missed their wives and children. Some suffered family and marital breakdowns because wives had extra-marital affairs or became addicted to gambling due to the long periods of loneliness while their husbands were at sea. For some fishers, working on a Taiwanese distant-water fishing vessel was no different from the traumatic experience of being incarcerated.

Improvements in shipbuilding technology and fishing techniques enabled Taiwanese fishers to explore more remote fishing grounds. Old, inefficient *Yakitama* engines were replaced by diesels that produced greater horsepower. Steel replaced wood as the main material of the vessel's hull and superstructure, which made the modular construction of large distant-water fishing vessels possible. The installation of large refrigerators could keep the catch fresh even on the long return voyage from the waters off northern Australia. Larger vessels also had bigger holds that enabled fishers to catch greater quantities on more distant fishing grounds. This further reduced operating costs, and the larger ratio of horsepower to tonnage helped the new vessels exploit remote fishing grounds more efficiently. These technological changes in shipbuilding had a positive impact on the development of the distant-water fishing industry in postwar Kaohsiung.

Both Taiwanese trawlermen and longlinermen fished in Southeast Asian waters, but the manner in which they diversified and exploited fishing grounds was very different. Trawlermen constantly had to look for new fishing areas because the marine ecosystems they exploited were prone to exhaustion. As a result, Taiwanese trawlermen abandoned the East China Sea, moved along the continental shelf to the Taiwan Strait and the South China Sea, and finally arrived in the waters off eastern Indonesia and northern Australia in the early 1970s. This expansion of Taiwanese trawl fishing grounds off Southeast Asia was "southward, one-way traffic;" no "U-turn" was allowed.

Taiwanese longline fishers in Southeast Asia, on the other hand, expanded their fishing grounds in a fan-shaped pattern; the eastern edge of their fishing territories was the Philippines archipelago, and the westernmost boundary was Hainan Island. The choice of longline fishing grounds depended upon seasonal conditions and/or market demand. Overseas supply bases had been established since the early 1960s, which enabled Taiwanese distant-water longliners to fish in all oceans. But the expansion into global-scale fisheries did not mean that the fishing grounds off Southeast Asia became less important. Investigations into tropical marine resources were still carried out by the government's experimental fishing vessels in the waters off Southeast Asia and northern Australia. A new wave of offshore longliners, owned mainly by Siao Liouciou immigrants and local Kaohsiung fishers, quickly filled the vacuum when longliners owned by local Kaohsiung tycoons left Southeast Asia in the 1960s to work in the Indian Ocean and the southwest Pacific.

Fishing on these remote grounds not only increased profits but also heightened the risk of being detained by foreign navies or marine police. Interdiction and detention occurred when Taiwanese fishers violated the territorial waters of coastal states, when they encountered typhoons or when they experienced engine malfunctions and had to enter a foreign port for repair. Detainment was a miserable experience; far worse, however, was that naval or police forces in Southeast Asia were likely to fabricate charges against the fishers. To avoid getting into such trouble, Taiwanese fishers sometimes bribed overseas marine law enforcement officers. This illegal practice, however, made some marine police even more corrupt and dangerous. For three decades Taiwanese fishing companies tried hard to arrange international fisheries cooperation arrangements with Southeast Asian countries, but they never succeeded before 1977. This was due to the fact that vessel owners did not have adequate knowledge of local practices and did not understand the proper procedures to carry out an international fishing cooperation agreement, especially without the assistance of the Taiwanese government. Thus, fraud and detainment frequently affected Taiwanese vessels. Moreover, Taiwan's fisheries authorities could not support the implementation of international fisheries cooperation arrangements or effectively protect the rights of vessel owners because Taiwan was diplomatically isolated due to Chinese political pressure.

In the 1970s, many serious problems simultaneously afflicted Taiwan's fishing industry. The rapid development of the island's industrial economy dramatically increased onshore job opportunities. Young men in fishing communities no longer had to make a living by going to sea. As well, wages in an onshore factory were no lower than what new fishers earned at sea. This employment situation not only discouraged a younger generation from working in the fishery but also drove experienced fishers to leave the industry. Even worse, over-fishing on the trawl-fishing grounds of Southeast Asia had become horrific. In 1977, Southeast Asian nations and smaller island states in the South Pacific, as well as many other countries, followed the United States and claimed a 200-mile EEZ to protect their coastlines and fisheries. Fishing zones in Southeast Asia and the South Pacific had been extremely crucial to the success of Kaohsiung's trawler and longline fisheries. Arranging cooperation agreements with other nations and investing more capital in exploiting new marine resources were two possible solutions to the problem. But the limited administrative and financial capability of the Taiwanese fishing companies prevented vessel owners from developing new fisheries or negotiating effectively with foreign governments.

As Taiwan's economy developed, the character of Kaohsiung's fishing communities changed. Cijin was always a much less-developed district which maintained the strongest sense of a closed fishing community. Gushan, on the other hand, was already urbanized in the prewar period. Fishers and their families lived in narrow alleys and flats and mingled with people engaged in onshore business activities. Gushan served throughout the postwar period as the centre of Taiwan's fishing industry. The post-1970s image of Gushan or Cijin as fishing communities was no longer as clear as in the early postwar years. After Cianjhen Fishing Port was built in the late 1960s, numerous fishing companies, ancillary industries and the government's key fisheries-related institutes shifted to Cianjhen, which now looked much like a modern industrial park rather than the home of an ocean-going fishing community.

In the 1970s, the radical development of Kaohsiung's onshore industries offered various new job opportunities. Women and children no longer needed to work in the ancillary industries. Moreover, young men were losing their drive to work at sea. As a solution to the growing labour shortage, foreign fishers were recruited in Southeast Asia to crew the vessels by the end of the 1980s. The appearance and character of Kaohsiung Fishing Port and the operations of the Kaohsiung distant-water fishing companies are now quite different than in earlier times. How the Taiwanese long-distance fishing companies manage and employ foreign fishers, and how the post-1980s Taiwanese fishing industry operates globally under the restrictions of 200-mile EEZs, are important topics for future research.

China started to develop its distant-water fishing industry in the late 1980s. With help of Taiwanese capital and fishing techniques, the Chinese industry has already made some significant achievements. Several important fishing ports with industrialized facilities have emerged along the Chinese coast. The diffusion of fishing techniques between Taiwan and China, and the developmental mode of the fishing industries in the two countries, also deserve further research, especially from the perspective of fishery policies and international politics.

Finally, the post-1970s labour shortage in Taiwan's fishing industry was solved through hiring fishers from neighbouring countries in Southeast Asia. The same problem of market forces, allocation and control in Japan and some European fishing nations occurred in the second half of the twentieth century, but for them the employment of foreign fishers was not a desirable option. How did they cope with such problems created by an acute labour shortage and high labour costs? What business strategies did Japan and Europe's fishing nations adopt? I believe that through comparative historical research into the Taiwanese, Japanese and European cases, historians and social scientists can gain a deeper understanding of the evolution and transformation of distant-water fishing in industrial countries at the beginning of the twenty-first century.

Appendix 1

List of Informants

Note: In Taiwan it is culturally incorrect for a young man to ask an older woman her name. Most of the elderly fishers felt offended when I asked to interview their wives. This is why all the names of the older fishers' wives are unknown.

Cai Bian (and his wife and female relatives)

Cai Bian (70+ yrs.) is a retired offshore fisher. He used to work in the aquaculture and coastal fishing industry in Cijin District, but he started to fish in the waters off Southeast Asia after the fishing port was built in Cijin. He is semi-illiterate, but his daughter is a university lecturer.

Cai Dingbang

Cai Dingbang (60+ yrs.) is Cai Wun'yu's nephew. He was elected as a supplementary National Assembly delegate by Kaohsiung's fishing communities. His fishing company still plays a leading role in Taiwan's distant-water fishing industry. He has worked as an honorary Consul-General for Papua New Guinea in Kaohsiung since 2003. He is active and well-known in Kaohsiung's political circles and fishing industry.

Cai Minghuei (and his wife)

Cai Minghuei (65+ yrs.) is a retired offshore fisher. He used to work in aquaculture and in the coastal fishing industry in Cijin District, but he started to fish in the waters of Southeast Asia after the fishing port was built in Cijin.

Cai Wun'yu

Cai Wun'yu (80+ yrs.) is a Kaohsiung business tycoon. He graduated from Keio University, one of Japan's best post-secondary institutions, in the colonial era. He was very active in both political circles and the fishing industry in postwar Taiwan. He served as the Head of the Kaohsiung Fishermen's Association and on the City Council. He was also invited to join the Taiwan Fisheries Production Committee. He was awarded an honorary doctorate of Marine

Science and Resources by Nihon University in Japan, and was elected to the Taiwan Provincial Assembly.

Chang Dejing

Chang (60+ yrs.) is a retired fishing master in Cianjhen. An indigenous Taiwanese, he worked in Kaohsiung's trawling industry for decades and used to fish in the waters off northern Australia.

Chen Cai

Chen Cai (75+ yrs.) was a successful fish wholesaler in the Kaohsiung Fish Market, but he switched careers and ran a successful fishing company.

Chen Jinjiao

Chen Jinjiao (75+ yrs.) is a retired fishing master who used to fish in the waters of Southeast China. He was still running a distant-water fishing fleet when I did my fieldwork in Kaohsiung.

Chen Jinyou

Chen Jinyou (about 50 yrs.) is a coastal fisher in Cijin who used to work in the distant-water fishing industry.

Chen Liousyu

Chen Liousyu (60+ yrs.) is an offshore longlinerman in Cijin who was still fishing in the waters off the Philippines and in the South China Sea when I was conducting my research in Kaohsiung.

Chen Shengli

Chen Shengli (65+ yrs.) began his fishing career in Siao Liouciou in his teens. He moved to Gushan in 1957 and launched his own offshore longliner fleet. He is still very active and well-known in Kaohsiung's Siao Liouciou fishing community.

Chen Shuteng (and his wife)

Chen Shuteng (60+ yrs.) is a Penghuan vessel owner. He moved to Gushan and started his fishing career in his teens. He worked at sea as a cook at the beginning of his career, but with Jhang Wunjhang's financial support he eventually started his own fishing company, which was still doing well when I was doing my fieldwork in Kaohsiung in 2002. He is semi-illiterate, but all his children have university degrees from Australia and the United States.

Chen Youyi (and his wife)

Chen Youyi (about 70 yrs.) is a retired Penghuan fishing master. He moved to Gushan and started his fishing career when he was a young man. He did not have a decent education because of his miserable childhood (Chen's father was killed at sea by an American bomber). Unlike some other fishers, Chen is cultured, refined and has a serene mind. He worked as a fishing master for both private and state-run fishing companies, and eventually became an important shareholder in a fishing enterprise.

Gu Tingfang

Gu (about 80 yrs.) was born in a wealthy fishing family in Shandong Province, where his family had been running a trawling business for generations. He fled to Keelung with his family when he was in his twenties. In the 1950s, he moved to Kaohsiung and worked as a manager for various fishing companies for many years, before starting his own fishing company in the early 1970s. He was very active and well-known in Kaohsiung's fishing industry. He was suffering from cancer when I did my fieldwork in Kaohsiung.

Guo Shihfu

Guo (65+ yrs.) is a retired fisher in Cijin District. He was running two offshore shrimp trawlers when I did my fieldwork in Kaohsiung.

Hong Fucai

Hong Fucai (about 70 yrs.) is a Siou Liouciou vessel owner who started to fish in his teens. He migrated to Gushan as a young adult and began his own offshore longliner company. He is still active in Kaohsiung's Siao Liouciou fishing community and is an elder in the Gushan Presbyterian Church.

Hong Shueijing

Hong Shueijing (about 70 yrs.) was born in a fishing family and used to fish in the waters off Southeast Asia. He was running a distant-water fishing fleet when I was in Kaohsiung in 2002.

Hong Sinjhu

Hong Sinjhu (70+ yrs.) is a Siou Liouciou vessel owner who started to fish as a teenager. He migrated to Takao in the prewar era and ran his own longliner fleet. Unlike other Siou Liouciou people, all his vessels work in the distant-water fishery.

Huang Fucyuan

Huang Fucyuan (75+ yrs.) is a retired fishing master. He received a fishery vocational education in the colonial period and worked in the distant-water fishing industry for decades.

Huang Sianchih

Huang (65+ yrs.) was head of the Fisheries Bureau. He received a MA in the US and was elected as supplementary National Assembly delegate by the fishing industry.

Huang Syue

Huang Syue (80+ yrs.) is a retired fishing master in Cijin and worked in Kaohsiung's offshore fishing industry for decades. His son is a fishing master as well.

Huang Yanshan

Huang Yanshan (about 70 yrs.) is the boss of a fishing company and is very active and well-known in Kaohsiung's fishing industry.

Jheng Sanbian

Jheng (70+ yrs.) is a retired Penghuan engineman. His mother died when he was very young and left him with several younger brothers. He moved to Gushan and started his fishing career after World War II. He worked as an enginemen for a long period and was detained by the Filipino authorities several times for illegal fishing. He is also a poet.

Jhong Ciourong

Jhong (60 + yrs.) is a retired fishing master.

Jin Nengjhen

Jin (70+ yrs.) is a retired fishing master and is originally from mainland China. He worked for the state-run fishing companies when he was young.

Li Jijhao

Li (75+ yrs.) is from Shandong. He ran a fishing fleet in Kaohsiung and was head of the Kaohsiung Fishing Boat Commercial Guild (KFBCG). He was very active in Kaohsiung's fishing industry. On behalf of the Guild, he negotiated on several occasions with the ancillary industries and played an important role in the negotiation of international fisheries cooperation arrangements.

Lin Changren

Lin (75+ yrs.) was born in Fujian Province, where he received a formal fishery education. He fled to Keelung in the 1940s and worked for private fishing companies as a fishing master. In the 1950s, he moved to Gushan and ran his own fishing companies. Unlike other mainland Chinese in Kaohsiung's fishing industry who stuck to trawling, he began by running pair-trawlers and then extended his business to longline and gill-net fishing. He was very well-known in Kaohsiung's fishing industry. One of his children migrated to Australia.

Liou Syuan

Liou (75+ yrs.) was a Penghuan vessel owner. He was an orphan, and in order to survive he moved to Kaohsiung and went to sea as a cook. He was soon promoted to fishing master. Liou tried to run his own fishing company, but he was unsuccessful and was forced to sell his vessels. He believed that his misfortune was a punishment for killing fish. He was suffering from dementia when I was in Kaohsiung in 2002. Although illiterate, he is also a poet.

Pan San'guang

Pan (60+ yrs.) was born into a wealthy local family and has a university degree. His family not only operated distant-water and offshore fishing fleets but also extended their business interests to ancillary industries like shipbuilding. They cooperated with American and Japanese fish dealers and pioneered global-scale fisheries in the 1960s. They are still powerful and influential in Kaohsiung.

Sia Fujhong

Sia (75+ yrs.) is a retired Penghuan fishing master but was born and grew up in Takao. He planned to join the Japanese navy during the war, but the conflict ended before he could enlist. He began his fishing career in the postwar era.

Su Liansan

Su (55+ yrs.) is a retired trawlerman.

Sun Bonan

Sun (70+ yrs.) is a trainer at Kaohsiung's fisheries training centre. He received a formal fishery education in the colonial period and worked as a fisher for a couple of years. He has trained numerous fishers for Taiwan's fishing industry.

Syu Yisin (and his wife)

Syu (55+ yrs.) is a retired Penghuan engineman. He studied at a fisheries school but did not finish the course. He worked in the distant-water fishing industry for a long period, but never as a fishing master. He was working at a shipyard in Cianjhen in 2002.

Wang Hejia

Wang (75+ yrs.) is a retired fisher. Unlike other older fishers, Wang went to senior high school. He worked in the offshore and distant-water fishing industries for decades. He used to work as a fishing master but was replaced because his fishing results were unsatisfactory. After he retired from fishing, he established a volleyball club and devoted himself to promoting the sport. With financial support from local enterprises and the Kaohsiung city government, several international games between Taiwan and Japan have been held.

Wu Dezan

Wu Dezan (60+ yrs.) is a fishing vessel owner. He has a university degree and is very active and well-known in Kaohsiung's fishing industry.

Wu Tianruei

Wu (70+ yrs.) was educated in the colonial period. He never went to sea, although he was born into a Kaohsiung fishing family. He ran a fish-processing factory and exported fish products to Japan. When I did my fieldwork in Kaohsiung in 2002, he was running a transportation fleet. He also lent money to other fishing companies. He is extremely wealthy and highly regarded in Kaohsiung's fishing industry.

Yu Jyun'yue

Yu (45+ yrs.) is the Chief Secretary of the KFBCG.

Zeng Jhengji

Zeng (60+ yrs.) is a retired Penghuan fishing master who moved to Kaohsiung to start his fishing career when he was a young boy. He worked in the distant-water fishing industry for decades. Unlike other fishers, he is a cultured person and can read some English.

Zhu Pingyi

Zhu (45+ yrs.) is a research fellow at the Academia Sinica. He used to work on Taiwan's maritime history but now is looking at cross-cultural scientific exchanges between China and Europe during the seventeenth and eighteenth centuries.

Zou Cinsheng

Zou (90+ yrs.) is from Shandong. After some time in the army, he started a fishing company and an ice-manufacturing plant. He was one of the very few Shandong people who still remained in the Kaohsiung fishing industry when I was in Kaohsiung in 2002. One of his children has migrated to Australia.

Select Bibliography

Sources in Chinese

The Agriculture and Forestry Industries in Taiwan [*Taiwan Nonglin*]. Nantou, 1946.

Annual Report of Effort and Catch Statistics by Area on Taiwan Demersal Fish Fisheries [*Taiwansheng Diyu Yuye Yuchangbie Yuhuo Tongji Nianbao*]. Taipei, 1970.

The Bluebook on US Aid and the Achievements of the Taiwanese Fishing Industry [*Taiwan Yuye Yunyong Meiyuan Chengguo Jiantao*]. Taipei, 1966.

A Brief Introduction to the Fishing Industry of Taiwan [*Taiwan Yuye Jianjie*]. Nantou, 1972.

A Brief Introduction to the Agricultural Industries, Vol. 10: Fisheries [*Nongye Yaolan Dishihji Yuyepian*]. Taipei, 1962.

A Brief Introduction to the Fishing Industry of Taiwan Province [*Taiwansheng Shueichan Gaikuang*]. Taipei, 1948.

The Bright Future of Taiwan's Fisheries [*Ciantu Yuanda De Taiwan Yuye*]. Taipei, 1955.

Chen, Hsien-Ming. "The Development of the Tuna Longline Fishery on Liuchiu Island" [*Lioucíouyu Jhih Weidiao Yuye Fajhan*]. *Geographical Research* [*Shihda Dili Yanjiou Baogao*], No. 33 (2000).

Chen, Wunshang, *et al.* (eds.). *The Ocean in Perspective* [*Haiyang De Jingsiang*]. Kaohsiung, 1999.

Chen, Zaifa, "Problems in the Development of Ocean Fisheries and their Solutions" [*Haiyang Yuye Fajhan Suo Mianlin De Wenti Ji Tantao*]. Unpublished KFBCG internal document, n.d.

China Fisheries Monthly [*Jhongguo Shueichan Yuekan*]. 1984.

China Fisheries News [*Jhongguo Yuye Sinwen*]. 1956-1974

Chinese Fisheries Weekly [*Jhonghua Yuye Zhoukan*]. 1975-1997.

The Construction of the Agriculture and Forestry Industries in Taiwan [*Taiwan De Nonglin Jianshe*]. Nantou, 1950.

The Development of the Fishing Industry [*Yuye Fajhan*]. Nantou, 1971.

Financial Survey of Taiwan's Fishery Enterprises of 1989 [*78 Nian Taiwan Dicyu Yuye Ciyeti Jingji Diaocha Baogao*]. Kaohsiung, 1991.

Fisheries Tribune [*Yuye Luntan*]. 1961-1974.

The Fishing Industry of Taiwan [*Taiwan Yuye*]. Nantou, 1953.

Fishing Ports of Kaohsiung City [*Gaosyongshih Yugang*]. Taipei, 2003.

Fu, Kuncheng. *A Study of the Foreign Courts' Decisions Regarding the Illegal Entry of Taiwan's Fishing Vessels* [*Waiguo Fayuan Jiou Woguo Yuchuan Feifa Rucin Anjian Suozuo Caipan Jhih Yanjiou Baogao Disan Jhih Yice*]. 2 vols. Taipei, 1990.

Hu, Singhua. *Chitchat about the Fisheries of Taiwan* [*Huayu Taiwan*]. Taipei, 2000 (online version).

______. *Maritime Taiwan* [*Haiyang Taiwan*]. Taipei, 2002 (online version).

______. *The Development of the Fishing Industry of Taiwan* [*Tuoyu Taiwan*]. Taipei, 1996 (online version).

Huang, Jyunjie (ed.). *Collection of Documents on JCRR* [*Jhongguo Nongcun Fucing Lianhe Weiyuanhuei Shihliao Hueibian*]. Taipei, 1991.

______. *The Reminiscences of the Staff Members of the Sino-American Joint Commission on Rural Reconstruction* [*Jhongguo Nongcun Fucing Lianhe Weiyuanhuei Koushu Lishih Fangwen Jilu*]. Taipei, 1992.

Huang, Sianchih. "The Past, Present, and the Future of Taiwan's Fisheries" [*Taiwan Yuye De Guocyu Sianzai Yu Weilai*]. Unpublished mss., n.d.

______. "Economic Development and the Structural Adjustment of the Production of Taiwan's Fisheries" [*Jingji Fajhan Yu Taiwan Yuye Shengchan Jiegou Jhih Tiaoshih*]. Unpublished paper presented to the Joint Annual Meeting of ROC Agricultural Groups [*Jhonghua Minguo Nongsyue Tuanti Lianhe Nianhuei Jhongsin Yiti Jhuti Baogao*], Taiwan, 1984.

Jhang, Baoshu. *Studies in the Establishment of China's Fisheries* [*Jhongguo Yuye Jianshe Yanjiou*]. Taipei, 1952.

Jhang, Sen. *The Management of Fisheries* [*Shueichan Jingyingsyue*]. Taipei, 1977.

Jhang, Shoujhen and Syu, Yinan (eds.). *The Past and Present of Hamasen* [*Hamasing De Cianshihjinsheng*]. Kaohsiung, 1998.

Jheng, Decing. *Looking Back on the History of Kaohsiung* [*Kanjian Laogaosyong*]. Kaohsiung, 2001.

Jhou, Sianwun. *The Economic History of Taiwan under the Japanese Empire, II* [*Rijyu Shihdai Taiwan Jingjishih II*]. Taipei, 1958.

Jhou, Siouhuan (ed.). *Documentary Collection on the Joint Commission on Rural Reconstruction, Vol. 1* [*Nongfuhuei Shihliao Diyijyuan*]. Taipei, 1995.

______ (ed.). *Joint Commission on Rural Reconstruction, Vol. 1: Government Developments* [*Nongfuhuei Shihliao Diyijyuan*]. Taipei, 1995.

Kaohsiung Fishing Boat Commercial Guild [*Gaosyong Yulun Shangye Tongye Gonghuei*]. Combined Meeting Records of the Directors and Supervisors [*Gaosyong Yulun Shangye Tongye Gonghuei Lijianshih Liansi Hueiyi Jilu*]. 1965-1980.

______. Conference Records of the Member Representatives [*Gaosyong Yulun Shangye Tongye Gonghuei Hueiyuan Daibiao Dahuei*]. 1965-1980.

______. Meeting Records of the Directors [*Gaosyong Yulun Shangye Tongye Gonghuei Lishih Hueiyi Jilu*]. 1965-1980.

Kaohsiung Pair-Trawler Owners Club [*Kaohsiung Shoucaowang Lianyihuei*]. Meeting Records [*Gaosyongshih Shoucaowangye Lianyihuei Hueiyi Jilu*]. 1962-1965.

Liou, Shaotang (ed.). "The Profiles of the Republic's Celebrities, 236" [*Minguo Renwu Siaojhuan* 236]. *Biographical Literature* [*Jhuanji Wensyue*], LXV, No. 6 (1994).

_____ (ed.). "The Profiles of the Republic's Celebrities, 140" [*Minguo Renwu Siaojhuan* 140]. *Biographical Literature* [*Jhuanji Wensyue*], XLIX, No. 3 (1986).

Naitō, Harukichi and Syu, Jiwu. *The History of the Fishing Industry in Taiwan* [*Taiwan Yuyeshih*]. Taipei, 1957.

Niou, Ke. "US Aid and the Economic Reform of Post-war Taiwan" [*Meiyuan Yu Zhanhou Taiwan De Jingji Gaizao*]. *American Studies Quarterly* [*Meiguo Yanjiu*], XVI, No. 3 (2002), http://www.edu.cn/20030718/3088422.shtml, accessed 9 April 2006.

"A Rambling Talk on Taiwan's Distant Water Pair-Trawl Fishing" [*Taiwan Yuanyang Shuangtuowang Yuye Mantan*]. Unpublished KFBCG internal document, n.d.

The References of the Ministry of Economic Affairs, Vol. 232 [*Jingjibu Cankao Zihliao*]. Taipei, 1961.

A Report on Taiwan's Distant-Water Longlining Fishing Industry [*Taiwancyu Yuanyang Weidiao Yuye Diaocha Baogao*]. Taipei, 1972.

Reports on Taiwan's Provincial Government [*Taiwansheng Shihjheng Baogao*]. Nantou, 1947.

Research on the Prices and Marketing Channels of Fish Products in Taiwan [*Taiwan Sianyu Yunsiao Yu Jiage Jhih Yanjiou*]. Taipei, 1956.

"The Running of Trawl Fishing" [*Tuowang Yuye Jingying Shihwu*]. Unpublished KFBCG internal document, n.d.

A Study on the Fishing Industry of Taiwan [*Taiwan Yuye Jhih Yanjiou*]. Taipei, 1974.

Sun, Tai'an. *The Distant Water Longline Fishing of Taiwan* [*Taiwan Yuanyang Weidiao Yuye*]. Taipei, 1973.

Survey on the Fishing Vessels and Fishing Gear of Taiwan [*Woguo Yuchuan Singneng Ji Yujyu Peihe Siankuang Jhih Diaocha Yanjiou*]. Kaohsiung, 1989.

Wang, Anyang and Liou, Sijiang. *The Problems of Taiwan's Distant Water Fisheries and Research on Fisheries Policies* [*Woguo Yuanyang Yuye Wenti Yu Jhengce Jhih Tantao*]. Taipei, 1979.

Wang, Taisheng. "The Constitutional Law of New Taiwan: Derivation from Pre-war Japan and Republican China" [*Taiwan Sianfa De Gushih: Cong 'Jiou Riben' Yu 'Jiou Jhongguo' Tueibian Ercheng 'Sin Taiwan'*]. *National Taiwan University Law Journal* [*Guoli Taiwan Dasyue Fasyue Luncong*], XXXII, No. 1 (2003), pp. 1-48.

Wu, Mingzuo. "A Brief Introduction to the Development of Taiwan's Trawl Fishing Industry" [*Taiwan Tuowang Yuye Fajhan Gaikuang*]. Unpublished internal document, KFBCG, n.d.

Wu, Tianren. *The Biographies of Those Who Engaged in the Early Fishing Industry of Taiwan* [*Taiwan Zaoci Yuye Renwujhih*]. Nantou, 1996.

Sources in Japanese

A Brief Guide to the Fishing Industry of Taiwan [*Taiwan Suisan Yōran*]. Taihoku, 1940.

A Brief Guide to the Fishing Industry of Takao [*Takao Shū Suisan Yōran*]. Takao, 1940.

A Brief History of Takao City, 10-Year Anniversary [*Takao Shisei Jū Shunen Ryakushi*]. Takao, 1934.

The Finance of Taiwan's Fishing Industry [*Taiwan Suisan Kinyū*]. Taihoku, 1930.

Fisheries Economics, Vol. 1: Fisheries Issues in the South. Tokyo, 1941.

The Fishing Industry of Taiwan [*Taiwan no Suisan*]. Taihoku, 1935.

The Fishing Industry of Takao [*Takao no Suisan*]. Takao, 1941.

Gotō, Kenichi. *Modern Japan and Southeast Asia* [*Kindai Nihon to Tōnan Ajia*]. Tokyo, 2001.

A Guide of Takao [*Takao Shū Yōran*]. Takao, 1941.

An Introduction to the Oceanic Fisheries of Our Country [*Honpō Kaiyo Gyogyō no Gensei*]. Tokyo, 1939.

Kamimura, Kendou. *Industries in Taiwan and Their Leading People* [*Taiwan Jigyōkai to Chūsin Jinbutsu*]. Taihoku, 1919.

Koshiyoshi, Yoshinobu. "The Fishing Industry of Taiwan" [*Taiwan no Suisangyō*]. In Takemoto, Iichirō (ed.). *Series on the Economy of Taiwan* [*Taiwan Keizai Sōsho*]. Taihoku, 1938, pp. 105-154.

______. "The Fishing Industry in Taiwan II" [*Taiwan no Suisangyō II*]. *Nanyo Suisan* [*Nanyo Fisheries*], III, No. 9 (1937), pp. 15-17.

Lin, Jinfa. *The History of the Development of Taiwan* [*Taiwan Hattatsu Shi*]. Taihoku, 1936.

Nakayama, Kaoru and Katayama, Kiyoo. *A Comprehensive Study of the Progress of Takao* [*Yakusin Takao no Zenbō*]. Takao, 1940.

Okamoto, Nobuo. *The History of Japan's Fishing Industry* [*Nihon Gyogyō Tsūshi*]. Tokyo, 1984.

The Present Situation of Taiwan's Fishing Industry [*Taiwan no Gyogyōgenkyō*]. Tokyo, 1978.

The Present Situation of Taiwan's Tuna Longline Fishing [*Taiwan ni okeru Maguro Gyogyō no Genkyō*]. Tokyo, 1981.

Tagame, Ichirou. *Takao Fishing Port and Its Onshore Facilities* [*Takao Gyokō to Sono Rikujō Setsubi*]. Takao, 1930.

Takahashi, Jun-ichi. *Japanese Whale Culture: Tracing the Development of Whale-hunting Culture* [*Kujira no Nihon Bunkashi: Hogei Bunka no Kōseki wo Tadoru*]. Kyoto, 1992.

Takemoto, Iichirō (ed.). *Series on the Economy of Taiwan 6* [*Taiwan Keizai Sōsho 6*]. Taihoku, 1938.

Yano, Tōru. *The Genealogy of Southward Development* [*Nanshin no Keifu*]. Tokyo, 1997.

Sources in English

Butcher, John G. "The Marine Fisheries of the Western Archipelago: Toward an Economic History, 1850 to the 1960s." In Pauly, Daniel and Martosubroto, Purwito (eds.). *Baseline Studies of Biodiversity: The Fish Resources of Western Indonesia*. Manila, 1996, pp. 24-39.

______. "Extending the Frontier: The Marine Fisheries of Southeast Asia since 1850." In Roch, Jean, *et al.* (eds.). *SOSEKIMA: Proceedings of Socio-economics, Innovation and Management of the Java Sea Pelagic Fisheries*. Jakarta, 1998.

______. "Getting into Trouble: The Diaspora of Thai Trawlers, 1975-2002." *International Journal of Maritime History*, XIV, No. 2 (2002), pp. 85-121.

______. *The Closing of the Frontier: A History of the Marine Fisheries of Southeast Asia, c. 1850-2000*. Singapore, 2004.

Chiu, Paul C.H. "Money and Financial Markets: The Domestic Perspective." In Ranis, Gustav (ed.). *Taiwan: From Developing to Mature Economy*. Boulder, CO, 1992, pp. 121-193.

Davies, Sam, *et al.* (eds). *Dock Workers: International Explorations in Comparative Labour History, 1790-1970*. 2 vols. Aldershot, 2000.

Gamba, Charles. *A Report on the Italian Fishermen of Fremantle*. Perth, 1952.

Gerrish, Margaret. "Following the Fish: Nineteenth Century Migration and Diffusion of Trawling." In Starkey, David J., *et al.* (eds.). *England's Sea Fisheries: The Commercial Sea Fisheries of England and Wales since 1300*. London, 2000, pp. 112-118.

Kalland, Arne. *Shingū: A Japanese Fishing Community*. London, 1981.

______ and Moeran, Brian. *Japanese Whaling, End of an Era*? London, 1992.

Pauly, Daniel and Martosubroto, Purwito (eds.). *Baseline Studies of Biodiversity: The Fish Resources of Western Indonesia*. Manila, 1996.

Ranis, Gustav (ed.). *Taiwan: From Developing to Mature Economy*. Boulder, CO, 1992.

Reeves, Peter, *et al.* "The Impact of 'Globalisation and Liberalisation' on South Asia's Fisheries: Experience in West Bengal and Bangladesh." *South Asia*, XXIV, No. 3 (2001).

Robinson, Robb. *Trawling: The Rise and Fall of the British Trawl Fishery*. Exeter, 1998.

Roch, Jean, *et al.* (eds.). *SOSEKIMA: Proceedings of Socio-economics, Innovation and Management of the Java Sea Pelagic Fisheries*. Jakarta, 1998.

Smith, M. Estellie (ed.). *Those Who Live from the Sea: A Study in Maritime Anthropology*. New York, 1977.

Starkey, David J., *et al.* (eds.). *England's Sea Fisheries: The Commercial Sea Fisheries of England and Wales since 1300*. London, 2000.

Taiwan's Fisheries. Nantou, 1971.

Thompson, Paul, *et al.* *Living the Fishing*. London, 1983.

Tull, Malcolm. "Fremantle, 1890-1990." In Davies, Sam, *et al.* (eds.). *Dock Workers: International Explorations in Comparative Labour History, 1790-1970*. 2 vols. Aldershot, 2000, II, pp. 471-493.

Tunstall, Jeremy, *The Fishermen*. London, 1962.

Websites

China College of Marine Technology and Commerce [*Jhongguo Haishih Shangye Jhuanke Syuesiao*], http://www.ccmtc.edu.tw/cjcmt-web/cjcmt/secretary/eng-ver/pagej.htm.

China Education and Research Network [*Zhongguo Jiaoyu He Keyan Jisuanji Wang*], http://www.edu.cn/20030718/3088422.shtml.

Chinese Warship Museum [*Jhongguo Jyunjian Bowuguan*], http://vm.rdb.nthu.edu.tw/cwm/taiwan/4202.html.

Defence Technical Information Centre, http://www.dtic.mil/doctrine/jel/doddict/data/i/02668.html.

ESCAP Virtual Conference, 1979-2006, http://www.unescap.org/DRPAD/VC/conference/ex_pi_17_ptf.htm.

FishBase, http://www.fishbase.org.

FishBase, http://fishbase.sinica.edu.tw.

Fisheries Information Service Site, http://www.fa.gov.tw/organization/sucess/success.ph.

Kaohsiung City Council, http://www.kcc.gov.tw/couser/s2.htm.

Kaohsiung Life Culture Education, http://www.kh.edu.tw.

Kaohsiung Museum of History, http://61.218.131.35/ks/old_photo/work.htm.

Lawphil Project, Arellano Law Foundation, http://www.lawphil.net.

Maritime Space: Maritime Zones and Maritime Delimitation, http://www.un.org/Depts/los/LEGISLATIONANDTREATIES.

Mie Prefectural Science and Technology Promotion Centre, http://www.mpstpc.pref.mie.jp/SUI/100aniv/20nantoumaru.htm.

My E Gov, http://www.gov.tw/EBOOKS/TWANNUAL/show_book.php?path=2_003_013.

News. yam. com, http:yam.udn.com/yamnews/daily/2912495.shtml.

Online Directory of Government Officials, http://twinfo.ncl.edu.tw/tiqry/hypage.cgi?HYPAGE=search/search_sim.hpg&dtd_id=5.

Private School Foundation, http://www.schoolfund.org.tw/List.

Sueichan Press, http://www.taiwan-fisheries.com.tw/news/2002/news-021204-6.htm.

Taiwan History Association], http://www.twhistory.org.tw/20010820.htm.

Taiwan Provincial Consultative Council, http://www.tpa.gov.tw.

Printed and bound by CPI Group (UK) Ltd, Croydon, CR0 4YY

16/09/2026

14963469-0003